OpenDaylight 構築実践ガイド

一般社団法人沖縄オープンラボラトリ
倉橋 良、鳥居 隆史、安座間 勇二、高橋 信行 =著

オープンソースSDN(Software Defined Network)
OpenDaylight&OpenStackで実現する
NFVオーケストレーション

インプレス

- 本書は、インプレスが運営するWebメディア「Think IT」で、「オープンソースのSDN OpenDaylightを始めよう！」として連載された技術解説記事を電子書籍およびオンデマンド書籍として再編集したものです。
- 本書の内容は、執筆時点までの情報を基に執筆されています。紹介したWebサイトやアプリケーション、サービスは変更される可能性があります。
- 本書の内容によって生じる、直接または間接被害について、著者ならびに弊社では、一切の責任を負いかねます。
- 本書中の会社名、製品名、サービス名などは、一般に各社の登録商標、または商標です。なお、本書では©、®、TMは明記していません。

目　次

第 1 章　イントロダクション「OpenDaylight とは何か？」　　1
1.1　OpenDaylight って何？　　3
1.2　OpenDaylight の特徴　　6
1.3　OpenDaylight を使うと何ができる？　　7

第 2 章　OpenStack with OpenDaylight（DevStack 編）　　9
2.1　環境・バージョンについて　　9
2.2　構築してみよう　　10
2.3　インスタンスを作ってみよう　　14
2.4　仕組みの解説　　16
2.5　SSH でインスタンスに接続したい　　28
2.6　ログの出力先　　29
2.7　（おまけ）シングルノードでの構築　　29

第 3 章　OpenStack with OpenDaylight（手動構築編）　　31
3.1　環境・バージョンについて　　31
3.2　構築してみよう　　32
3.3　tcpdump を使ってデバッグするポイント　　35
3.4　（おまけ）シングルノードでの構築　　37

第 4 章　RESTCONF API を使ってフローを書き換えてみる　39

- 4.1　RESTCONF API　39
- 4.2　OpenDaylight RESTCONF API Documentation　39
- 4.3　RESTCONF API を使ってみる　40

第 5 章　OpenStack Tacker による NFV オーケストレーション　51

- 5.1　NFV アーキテクチャのおさらい　51
- 5.2　環境・バージョンについて　52
- 5.3　構築してみよう　53
- 5.4　Tacker で VNF をデプロイしてみよう　55
- 5.5　VNF インスタンスは service テナントに作られる　58
- 5.6　VNF の構成　58
- 5.7　VNF を使ってみる　59

第 6 章　OpenDaylight でクラスタを組んでみよう　69

- 6.1　OpenDaylight のデータストアとクラスタ　69
- 6.2　Raft　71
- 6.3　環境・バージョンについて　71
- 6.4　構築してみよう　73
- 6.5　動作確認　80

付録 A　Service Function Chaining　89

第1章 イントロダクション「OpenDaylight とは何か？」

近頃、「OpenDaylight」という単語の露出が日本のメディアでも増えてきています。何かしらの製品のベースとして使わるようにもなってきました。ですがその一方で、OpenDaylight を技術的にわかりやすく解説したドキュメントがなく、インストールしてもどう使うかわからない！ と困っている方もいると思います。沖縄オープンラボラトリ (http://www.okinawaopenlabs.org/) では約 2 年前から実際に OpenDaylight をさわって、いくつかのシステム検証を行ってきています。本書では、現場で OpenDaylight をいじり倒してきたメンバーが、OpenDaylight の簡単な始め方についてわかりやすく伝授します。

SDN（Software-Defined Networking）を知っている方は多くなってきたと思います。各ベンダーからは VMware NSX、NEC ProgrammableFlow、Juniper Contrail のような SDN 技術/製品が提案、販売されています。ですが「OpenDaylight」となると、名前を聞いたことがある、というくらいではないでしょうか？ 日本では、そこそこ SDN を知っていたり、OpenFlow に詳しい業界の方であっても、「OpenDaylight は難しい」「よくわからない」「まだまともには動かないらしい」という情報が出回っているように思えます。

一方、海外での反応はというと、OpenDaylight はすでに「成功を約束されたオープンソース」と言われています。それは OpenDaylight がコミュニティとして大きくなってきていること、ベンダーニュートラルに運営されガバナンスもされていること、コードが改善されてきており商用でも使えるレベルに達していること、などが要因として挙げられます。ところが国内では日本語の情報が少ないためか、なかなか広まらないのが現状です。ベンダーが集まって開発しているため、ベンダー主導で、オープンなコミュニティではないのでは？ という見方もあるようです。ベンダーの人が開発しているのは正しいのですが、コミュニティはオープンに運営されています。特定のベンダーが牛耳っていたり、方向性を決めているということはありません。

第 1 章　イントロダクション「OpenDaylight とは何か？」

　ところで、オープンソースの SDN コントローラや OpenFlow コントローラというと、古くは NOX や Floodlight、国内では Trema や Ryu、最近では ONOS、商用からオープンソースにした MidoNet など、いろいろあります。その中で OpenDaylight はどう位置付けられるのか、ほかのツールとどこが違うのか、という疑問がでてくるでしょう。これを知るためにまず OpenDaylight が生まれた時期を振り返ってみましょう。

　OpenDaylight プロジェクトは 2013 年 4 月に Linux Foundation によって発足しました。この前年、Nicira が VMware に買収されています（2012 年 7 月）。Cisco は Cisco ONE というコンセプトで SDN の取り組みを強化していました。OpenFlow から SDN へ、スタートアップから大手へと流れが出てきた時期でした。ベンダー各社は、SDN、OpenFlow という技術をどうマーケットに出していき、顧客をつかんでいくか、を真剣に考え始めました。そこで重要なのは、ユースケースであり、アプリケーション・システムとどう連携をして価値を生み出すかでした（これは今も変わりませんが）。そのために各社の共通部分、例えば OpenFlow コントローラは各社が別々に開発をするのではなく、非競争領域として共通的なものをつくり、そのうえでのアプリケーションや API の使い方で各社が競争をしよう、という考え方が出てきました。もちろんオープンソースの OpenFlow コントローラは当時からありました。ですが、より汎用的で、各社が使いやすいものが必要とされました。OpenFlow だけでは装置の設定（Configuration）はやりきれません。ポートの設定や VLAN の設定が必要な場合もあります。BGP（Border Gateway Protocol）などのルーティングプロトコルの設定も必要です。それらを網羅できるもの、つまりプラットフォームを作ろう、という目的で OpenDaylight プロジェクトが始まりました。OpenDaylight は SDN コントローラではなくプラットフォームだ、と言っているのはそういう背景からです。

　このように、OpenDaylight はそもそもベンダーが商用として使うことを目的として開発されています。オープンソースというと、実験的なものであったり、ソフトウェア開発者が企業の枠を超えて作りたいものを作るというイメージがありますので、OpenDaylight は少し毛色が違うように見えるかもしれません。ですが、ベンダーが注力をしてきたからこそ、短期間で様々な機能を実装し、安定化も進んできています。ベンダー主導をネガティブに捉えるのではなく、SDN を実現する選択肢の一つとして評価・検討していくべきだろうと思います。

　本書では、OpenDaylight を使ってみることにフォーカスをし、まずは OpenStack との組み合わせで実際にどう動かすかを解説します。OpenDaylight はプラットフォームなので、ただダウンロードして起動すれば動くというものではありません。システムの中に組み入れて、初めて動かすことができます。現時点では、それを一番簡単に試すことができるのが、OpenStack との組み合わせです。しかも OpenStack には DevStack というツールが用意されており、それ

を使えばダウンロードからインストールまでを自動で行ってくれます。次回からOpenStack +
OpenDaylightの構築について詳しく説明をします。ご期待ください。

　それに先立ち、本章ではまずOpenDaylightについての基本的な情報をお伝えします。

1.1　OpenDaylightって何？

　すでに触れたように、OpenDaylightプロジェクトは2013年4月にLinux Foundationに
よって発足しました。主要ネットワーク関連ベンダーが開発に参加しており、2014年2月に最
初のバージョンであるHydrogenがリリースされました。OpenDaylightの各バージョンの名前
は化学元素名が使われており、Hydrogenは元素番号1番目の「水素」を意味します。つい先日
（2016年2月22日）に最新バージョンであるBerylliumがリリースされました。

　次のグラフは主要な貢献者数とコミット数を他のSDNコントローラと比較したものです。こ
れを見ると、全体の半分以上をOpenDaylightが占めており、SDNコントローラの中でも特に
高い注目や期待をされていることが分かります。コミュニティも活発であり、2015年7月27日
〜31日にOpenDaylight Summitがアメリカのサンタクララで開催されました。Keynoteでは
各社のOpenDaylightに対する取り組みやOpenDaylightのExecutive DirectorであるNeela
Jacques氏によるOpenDaylightの成長、NFV領域への積極的な参画といった紹介がありま
した。

第1章 イントロダクション「OpenDaylight とは何か？」

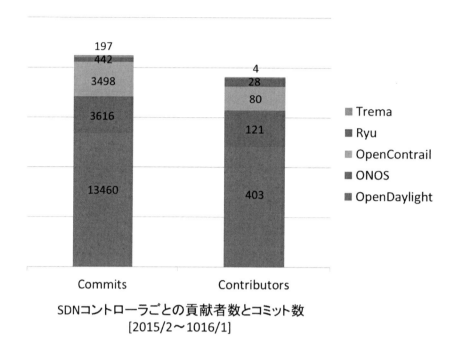

図1.1　https://www.openhub.net/ より

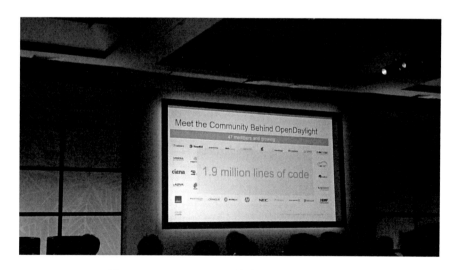

図1.2　OpenDaylight の規模の大きさが伺える

1.1 OpenDaylight って何？

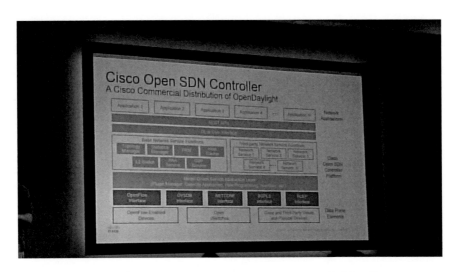

図 1.3　各社の ODL ベースの SDN コントローラ（Cisco）

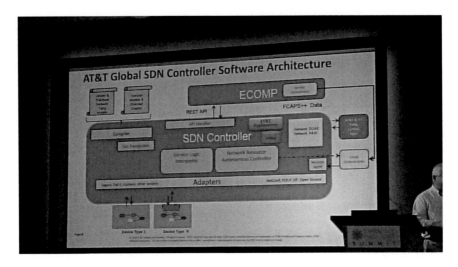

図 1.4　各社の ODL ベースの SDN コントローラ（AT&T）

OpenDaylightには地域ごとにユーザーグループと呼ばれるコミュニティが結成されています。日本でも「OpenDaylight Tokyo User Group」の名称で定期的にハンズオンなどが開催され、積極的に活動しています。(http://www.meetup.com/OpenDaylight-Tokyo-User-Group/)

1.2 OpenDaylightの特徴

OpenDaylightは様々なSouthbound Protocol（OpenFlow、NETCONF、BGPなど）をサポートしています。そのため、レガシーネットワーク、オーバーレイネットワーク、OpenFlowネットワークといった、様々なネットワークを一元的に管理、制御することが可能になります。また、豊富なアプリケーションも備えており、OpenFlowを使ったHop-by-Hopにおけるマルチテナントのネットワークを作成するVTN（Virtual Tenant Network）、SNMPを使って機器を管理・制御するSNMP4SDNやNFVを実現するためのSFC（Service Function Chaining）などがあります。

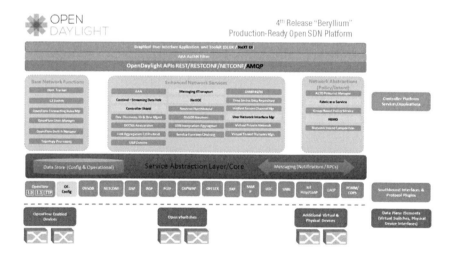

図1.5 OpenDaylight（Beryllium）ブロックダイアグラム（https://www.opendaylight.org/odlbe）

実装言語はJava（一部Java以外の言語も含む）で、Javaの開発経験がある方であれば馴染みのある、Eclipseで採用されているOSGIフレームワークを基盤として実装されています。OSGIの実行環境はApache Karafを使用しており、各モジュールはPluginとして実装されているため、OpenDaylightのサービスを停止することなく、かつ他の機能に影響がでないようモジュールの追加や削除が可能です。

1.3　OpenDaylight を使うと何ができる？

　OpenDaylight は多くの Southbound Protocol をサポートしているので、データセンター内でのオーバーレイネットワークの管理・制御、OpenFlow を使ったマルチテナンシーなネットワークの構築など、多くのベンダーで検証が始められています。

　AT&T はインターネットサービスや映像配信サービスなどを行っているアメリカ最大手の電話会社です。AT&T は OpenDaylight ベースの独自グローバルコントローラを開発し、従来のキャリアが管理する L3 までの枠を超え、アプリケーション層（L4-L7）までを制御することに挑戦しています。

　KT Corporation は韓国における最大級の通信事業者です。KT では広域ネットワーク網における、各ネットワーク機器が持つ EMS（Element Management Systems）を、従来は EMS ごとに管理が必要でしたが、それらを一元的に中央管理するためのトランスポート SDN コントローラとして OpenDaylight を採用しています。

　このように、OpenDaylight はベンダーだけでなくユーザーも巻き込んで開発が進んでいます。今回リリースされた Beryllium あたりからは、国内でも商用で使おうという話がでてくると期待されています。本書では、これからホットになるであろう OpenDaylight を使ってみるための情報を提供していきます。

第2章 OpenStack with OpenDaylight（DevStack編）

　本章は、DevStack（http://docs.openstack.org/developer/devstack/ ）というツールを使っ
てOpenDaylight（以降、ODLと省略）とOpenStackを連携させた環境の構築にチャレンジし
てみたいと思います。

　それではさっそく構築方法について説明していきたいのですが、いきなり「ODLとOpenStack
をそれぞれマニュアルに沿って構築して連携させてみましょう！」、というのは少し難易度が高
いです。というのも、ODLは比較的新しめのOSSであるために情報が雑多で探しづらく、日本
語の情報も少ないというのが現状だからです。

　そこで今回は、もっとも簡単とされるDevStackを使った構築方法をご紹介していきます。
DevStackはOpenStack開発者達の間では、公式のテスト環境構築ツールとして親しまれてい
ます。数十行の設定ファイルを書いてスクリプトを実行すれば簡単にOpenStack環境ができて
しまいます。さらに、そのDevStackの設定ファイルにODL用のコンフィグレーションを書き
足せば、ODLとOpenStackの連携も簡単にできてしまうのです。

2.1 環境・バージョンについて

　今回の構築はマルチノード構成なので2台のVMを使用します。筆者の環境ではvSphere
ESXi上に2台のUbuntuのVMを用意しています。この2台のVMは、DevStackの性質上、
ルート権限（sudo）での実行が可能である必要があります。さらに、ソースコードのダウンロー
ドを行うためインターネットアクセスができる環境も必要です。

第 2 章　OpenStack with OpenDaylight（DevStack 編）

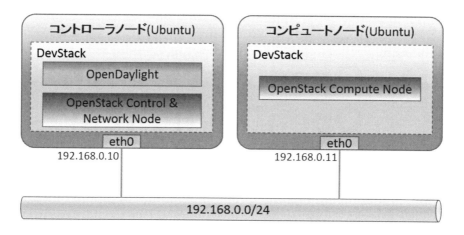

図 2.1　ノード構成図

OS ディストリビューション	カーネルバージョン
Ubuntu 14.04.3 LTS	Linux version 3.13.0-74-generic

　OpenStack 環境の各コンポーネントは 2015 年 10 月にリリースされた Liberty を使用します。

コンポーネント	バージョン
Nova	
Glance	
Neutron	
Horizon	
networking-odl	Liberty

　今回の構築では、ODL は DevStack が自動構築することになりますが、特に指定が無ければ Lithium リリースが選択されます。

OpenDaylight リリース	ビルドバージョン
Lithium	lithium-snapshot-0.3.5

2.2　構築してみよう

全ノード共通の設定

　それでは早速構築を始めていきましょう。まず、全ノードで共通の設定を先にやっておきます。先にファイアーウォールは無効化しておきます。

2.2 構築してみよう

```
~$ sudo ufw disable
```

次に DevStack をダウンロードしてきます。その際、git コマンドが必要なので入っていなければ適宜インストールしてください。ブランチ名は stable/liberty を指定します。※ 2016 年 11 月 17 日を過ぎると liberty-eol(End-of-Life) タグを指定しなければならなくなるかもしれません。

```
~$ sudo apt-get install git
~$ git clone --branch stable/liberty https://github.com/openstack-dev/devstack.git
~$ cd devstack/
```

コントローラノードの構築

続いて、コントローラノードの構築に入ります。コントローラノードの local.conf のサンプルは、下記 URL よりダウンロードできます。local.conf は devstack/配下に置きます。

https://github.com/YujiAzama/opendaylight-openstack-integration/blob/master/allinone/multi-node/control/local.conf

ダウンロードした local.conf をテキストエディタで開くと以下の様になっています。まず、お使いの環境に合わせて HOST_IP をコントローラノードの IP アドレスに変更してください。

```
 :
# IP Details
HOST_IP=192.168.0.10
SERVICE_HOST=$HOST_IP
 :
# Neutron
 :
enable_plugin networking-odl http://git.openstack.org/openstack/networking-odl
${BRANCH_NAME}

# OpenDaylight Details
ODL_MODE=allinone
ODL_PORT=8181
```

ODL に関する最低限必要な設定は、networking-odl の有効化（enable_plugin）、動作モード（ODL_MODE）の設定、ポート番号（ODL_PORT）の設定の 3 つです。

- networking-odl とは ODL のための Neutron ML2 メカニズムドライバの外部ライブラリの事です。
- 動作モードは DevStack が ODL を構成するために必要となります。
- ポート番号は ODL のポート番号ですが、ODL のデフォルトである 8181 番ポートを指定

第 2 章　OpenStack with OpenDaylight（DevStack 編）

します。

　動作モードは ODL に関する enable_service の設定を簡単にしてくれます。動作モードとして ODL_MODE に指定できるものを以下の表に示します。

動作モード	説明
allinone	この DevStack インスタンスで ODL を実行したい場合に指定する。シングルまたはマルチノードの場合に有用
externalodl	DevStack で ODL を管理しない。開発環境用に使用される
compute	DevStack インスタンスが Compute ノードの場合に指定する
manual	有効にするサービスを明示的に指定する

　今回、ODL は OpenStack コントローラノードと同じホストで実行させたいので allinone を指定します。

　もし manual を指定する場合、それぞれの動作モードと同じ動作をさせるためには以下の様に設定します。

- allinone と同じ動作

```
[[local|localrc]]
enable_sevice odl-server
enable_sevice odl-compute
```

- externalodl と同じ動作

```
[[local|localrc]]
enable_sevice odl-neutron
enable_sevice odl-compute
```

- compute と同じ動作

```
[[local|localrc]]
enable_sevice odl-compute
```

　これらの設定ができれば、あとは stack.sh スクリプトを実行するだけです。スクリプトが終了するまでに十数分程度かかる場合がありますので気長に待ちましょう。以下の様なメッセージが出力されれば終了です。

```
~/devstack$ ./stack.sh
  :
This is your host IP address: 192.168.0.10
This is your host IPv6 address: ::1
Horizon is now available at http://192.168.0.10/dashboard
Keystone is serving at http://192.168.0.10:5000/
The default users are: admin and demo
The password: password
~/devstack$
```

以上でコントローラノードの構築は完了です。

コンピュートノードの構築

続いて、コンピュートノードの構築です。コントローラノードと同様に、下記 URL より local.conf のサンプルをダウンロードします。

https://github.com/YujiAzama/opendaylight-openstack-integration/blob/master/allinone/multi-node/compute/local.conf

ダウンロードした local.conf を開いて、HOST_IP をコンピュートノードの IP アドレスへ、SERVICE_HOST をコントローラノードの IP アドレスへ変更してください。動作モードの設定である ODL_MODE は compute を指定します。

```
~/devstack$ vim local.conf
  :
# IP Details
HOST_IP=192.168.0.11
SERVICE_HOST=192.168.0.10
RABBIT_HOST=$SERVICE_HOST

# Neutron
  :
enable_plugin networking-odl http://git.openstack.org/openstack/networking-odl stable/liberty

# OpenDaylight
ODL_MODE=compute
ODL_PORT=8181
```

あとはコントローラノードと同様に、stack.sh スクリプトを実行するだけです。以下の様なメッセージが出力されれば終了です。

```
~/devstack$ ./stack.sh
  :
This is your host IP address: 192.168.0.11
This is your host IPv6 address: ::1
```

```
2016-01-13 07:21:51.305 | stack.sh completed in 56 seconds.
~/devstack$
```

以上で構築作業は終了です。たったこれだけで、ODL と OpenStack が連携した環境の構築ができてしまいました。

2.3　インスタンスを作ってみよう

構築ができていろいろ動かしてみたくなったと思います。さっそくネットワークやインスタンスを作ってみましょう。コントローラノードのコマンドラインからオペレーションしていきます。

```
~/devstack$ source openrc demo demo                                          # 認証用
の環境変数を設定（demo ユーザー、demo テナント）
~/devstack$ neutron net-create net01                                         # ネット
ワークを作成
~/devstack$ neutron subnet-create --name net01-subnet net01 10.11.12.0/24    # サブ
ネットを作成
~/devstack$ nova boot --image cirros-0.3.4-x86_64-uec --flavor m1.tiny vm01  # インス
タンス（vm01）を作成
~/devstack$ nova boot --image cirros-0.3.4-x86_64-uec --flavor m1.tiny vm02  # インス
タンス（vm02）を作成
```

作成したネットワークとインスタンスを OpenStack の管理画面である Horizon の GUI で確認してみましょう。ブラウザから以下の URL にアクセスするとログイン画面が表示されます。
※ IP アドレスはコントローラノードの IP アドレスです。

　http://192.168.0.10/dashboard

ユーザー名に「demo」、パスワードに「password」を入力して接続します（図 2.2）。

ネットワークタブのネットワークトポロジを選択すると、先ほど作成したネットワークとインスタンスが確認できます（図 2.3）。

2.3 インスタンスを作ってみよう

図 2.2　Login

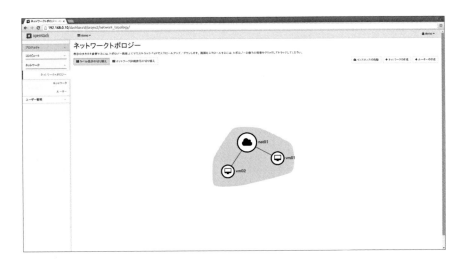

図 2.3　Topology

続いて ODL の GUI も見てみましょう。ブラウザから以下の URL にアクセスすると OpenDaylight User Experience(DLUX) のログイン画面が表示されます。※ IP アドレスはコントローラノードの IP アドレスです。

http://192.168.0.10:8181/index.html

ユーザー名に「admin」、パスワードに「admin」を入力してログインします（図 2.4）。

図 2.4　Login

2 台のスイッチが接続されているネットワークトポロジが確認できます。それぞれコントローラノードとコンピュートノードのブリッジを示しています（図 2.5）。

2.4　仕組みの解説

構築ができたので、ODL と OpenStack が連携する仕組みを見ていきましょう。

まず、内部のネットワークを説明します。全てのノードに br-int という Open vSwitch（以降、OVS と省略）のブリッジが作成されます。ネットワークとサブネットを作成するとコントローラノードの br-int に DHCP サーバのポートが作成されます。インスタンスを作成するとコンピュートノードの br-int に TAP デバイスとして接続されます。ノード間は VxLAN でトンネリングされているので、インスタンスからの DHCP リクエストはそのトンネルを通って DHCP Agent に届きます（図 2.6）。

次に、OpenStack Neutron が ODL を使ってどのように各ノードの OVS に対してフローを

2.4 仕組みの解説

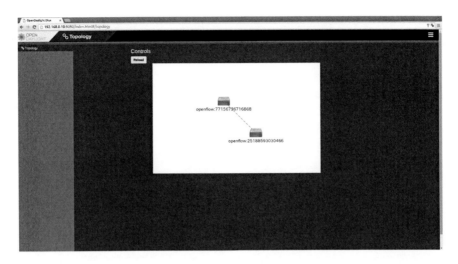

図 2.5　Topology

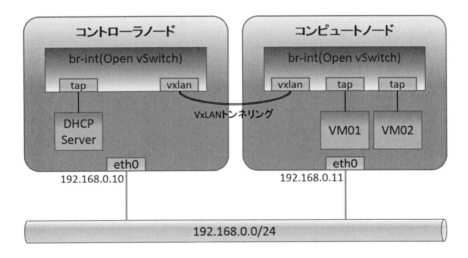

図 2.6　Internal bridges

設定するのかを見ていきます。OpenStack Neutron は、ML2 プラグインのメカニズムドライバとして用意されている ODL のドライバを経由し、ODL に対してポート作成などのリクエストを発行します。ODL はそれらのリクエストを Neutron NBI(North Bound Interface) から受けとって MD-SAL に渡します。

MD-SAL(Model-Driven Service Abstraction Layer) とは、North Bound API と South Bound API の間でやりとりされる様々なデータを統一する抽象化フレームワークです。

17

第 2 章　OpenStack with OpenDaylight（DevStack 編）

データ抽象化のモデリング言語として YANG が採用されています。MD-SAL は OVSDB と OF-Plugin の South Bound API を使ってコンピュートノードの OVS のフローを制御します（図 2.7）。

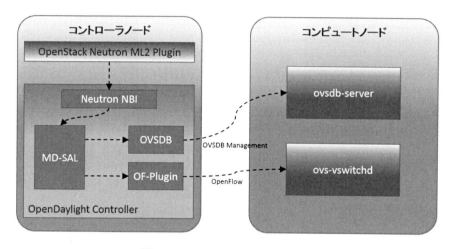

図 2.7　OpenDaylight controller

ではここからは実際に OVS のブリッジやフローがどのような状態になっているのか見ていきましょう。構築直後の状態とネットワークとサブネットを作成した状態、インスタンスを起動した状態で、ブリッジのポートやフローにどのような設定がされているかを見ていきます。

まず、構築直後の OVS のブリッジとポートの状態を確認してみます。Manager にコントローラノードの IP アドレスが設定されていることから OVSDB-Server のマネージャーに ODL が使われている事がわかります。さらに、br-int ブリッジが作成され、Controller にコントローラノードの IP アドレスが設定されていることから、ODL が OpenFlow コントローラーとして使われている事も確認できます。

[コントローラノード]

```
~/devstack$ sudo ovs-vsctl show
43cbe79e-8e51-4a5a-8638-03df224ad777
    Manager "tcp:192.168.0.10:6640"  # コントローラノードの IP アドレスが指定される
        is_connected: true
    Bridge br-int  # br-int ブリッジが作成される
        Controller "tcp:192.168.0.10:6653"  # コントローラノードの IP アドレスが指定される
            is_connected: true
        fail_mode: secure
        Port br-int
            Interface br-int
```

2.4 仕組みの解説

```
            type: internal
    Bridge br-ex
        Port br-ex
            Interface br-ex
                type: internal
    ovs_version: "2.0.2"
```

[コンピュートノード]

```
~/devstack$ sudo ovs-vsctl show
00790da7-431d-430c-b638-9114e2f17552
    Manager "tcp:192.168.0.10:6640"   # コントローラノードの IP アドレスが指定される
        is_connected: true
    Bridge br-int   # br-int ブリッジが作成される
        Controller "tcp:192.168.0.10:6653"   # コントローラノードの IP アドレスが指定される
            is_connected: true
        fail_mode: secure
        Port br-int
            Interface br-int
                type: internal
    ovs_version: "2.0.2"
```

この時の各ノードの初期状態のフローを確認してみると、全てのパケットが110番テーブルで破棄されるようになっています。

[コントローラノード]

```
~/devstack$ sudo ovs-ofctl dump-flows br-int -O OpenFlow13   # フローテーブルの一覧を取得
OFPST_FLOW reply (OF1.3) (xid=0x2):
 cookie=0x0, duration=1735.11s, table=0, n_packets=0, n_bytes=0, dl_type=0x88cc
actions=CONTROLLER:65535
 cookie=0x0, duration=1733.976s, table=0, n_packets=2, n_bytes=140, priority=0
actions=goto_table:20
 cookie=0x0, duration=1733.969s, table=20, n_packets=2, n_bytes=140, priority=0
actions=goto_table:30
 cookie=0x0, duration=1733.966s, table=30, n_packets=2, n_bytes=140, priority=0
actions=goto_table:40
 cookie=0x0, duration=1733.961s, table=40, n_packets=2, n_bytes=140, priority=0
actions=goto_table:50
 cookie=0x0, duration=1733.955s, table=50, n_packets=2, n_bytes=140, priority=0
actions=goto_table:60
 cookie=0x0, duration=1733.953s, table=60, n_packets=2, n_bytes=140, priority=0
actions=goto_table:70
 cookie=0x0, duration=1733.949s, table=70, n_packets=2, n_bytes=140, priority=0
actions=goto_table:80
 cookie=0x0, duration=1733.944s, table=80, n_packets=2, n_bytes=140, priority=0
actions=goto_table:90
 cookie=0x0, duration=1733.939s, table=90, n_packets=2, n_bytes=140, priority=0
actions=goto_table:100
 cookie=0x0, duration=1733.935s, table=100, n_packets=2, n_bytes=140, priority=0
actions=goto_table:110
```

第 2 章　OpenStack with OpenDaylight（DevStack 編）

```
 cookie=0x0, duration=1733.902s, table=110, n_packets=2, n_bytes=140, priority=0
actions=drop
```

［コンピュートノード］

```
~/devstack$ sudo ovs-ofctl dump-flows br-int -O OpenFlow13   # フローテーブルの一覧を取得
OFPST_FLOW reply (OF1.3) (xid=0x2):
 cookie=0x0, duration=1242.665s, table=0, n_packets=0, n_bytes=0, dl_type=0x88cc
actions=CONTROLLER:65535
 cookie=0x0, duration=1240.786s, table=0, n_packets=4, n_bytes=320, priority=0
actions=goto_table:20
 cookie=0x0, duration=1240.783s, table=20, n_packets=3, n_bytes=230, priority=0
actions=goto_table:30
 cookie=0x0, duration=1240.781s, table=30, n_packets=3, n_bytes=230, priority=0
actions=goto_table:40
 cookie=0x0, duration=1240.781s, table=40, n_packets=3, n_bytes=230, priority=0
actions=goto_table:50
 cookie=0x0, duration=1240.775s, table=50, n_packets=3, n_bytes=230, priority=0
actions=goto_table:60
 cookie=0x0, duration=1240.77s, table=60, n_packets=3, n_bytes=230, priority=0
actions=goto_table:70
 cookie=0x0, duration=1240.764s, table=70, n_packets=3, n_bytes=230, priority=0
actions=goto_table:80
 cookie=0x0, duration=1240.753s, table=80, n_packets=3, n_bytes=230, priority=0
actions=goto_table:90
 cookie=0x0, duration=1240.743s, table=90, n_packets=3, n_bytes=230, priority=0
actions=goto_table:100
 cookie=0x0, duration=1240.737s, table=100, n_packets=3, n_bytes=230, priority=0
actions=goto_table:110
 cookie=0x0, duration=1240.73s, table=110, n_packets=3, n_bytes=230, priority=0
actions=drop
```

次に、ネットワークとサブネットが作成された状態のブリッジとポートの状態を見てみます。各ノードの br-int に、ノード間接続のための VxLAN のポートが追加されているのが確認できます。さらに、コントローラノードの br-int に tap から始まる名前のデバイスが追加されていると思います。これは DHCP サーバのポートです。

［コントローラノード］

```
~/devstack$ sudo ovs-vsctl show
43cbe79e-8e51-4a5a-8638-03df224ad777
    Manager "tcp:192.168.0.10:6640"
        is_connected: true
    Bridge br-int
        Controller "tcp:192.168.0.10:6653"
            is_connected: true
        fail_mode: secure
        Port "vxlan-192.168.0.11"   # コンピュートノードとのトンネリング用ポート
            Interface "vxlan-192.168.0.11"
                type: vxlan   # トンネリングのタイプは VxLAN
```

```
            options: {key=flow, local_ip="192.168.0.10",
remote_ip="192.168.0.11"}
        Port br-int
            Interface br-int
                type: internal
        Port "tap178a8b99-70"   # DHCPサーバのポート
            Interface "tap178a8b99-70"
                type: internal
    Bridge br-ex
        Port br-ex
            Interface br-ex
                type: internal
    ovs_version: "2.0.2"
```

[コンピュートノード]

```
~/devstack$ sudo ovs-vsctl show
00790da7-431d-430c-b638-9114e2f17552
    Manager "tcp:192.168.0.10:6640"
        is_connected: true
    Bridge br-int
        Controller "tcp:192.168.0.10:6653"
            is_connected: true
        fail_mode: secure
        Port br-int
            Interface br-int
                type: internal
        Port "vxlan-192.168.0.10"   # コントローラノードとのトンネリング用ポート
            Interface "vxlan-192.168.0.10"
                type: vxlan    # トンネリングのタイプはVxLAN
                options: {key=flow, local_ip="192.168.0.11",
remote_ip="192.168.0.10"}
    ovs_version: "2.0.2"
```

tapから始まるデバイス名はNeutronのポートIDの先頭11桁と対応しているので、neutronコマンドを使ってDHCPのポートが追加されたことが確認できます。

```
~/devstack$ neutron port-list
+--------------------------------------+------+-------------------+------------------
| id                                   | name | mac_address       | fixed_ips
|
+--------------------------------------+------+-------------------+------------------
| 178a8b99-700f-4d2f-be35-f34d420f66a3 |      | fa:16:3e:ef:85:07 | {"subnet_id":
"8d373fbd-6134-47d1-aff6-86ee1ae7c8d1", "ip_address": "10.11.12.2"} |
+--------------------------------------+------+-------------------+------------------
~/devstack$ neutron port-show 178a8b99-700f-4d2f-be35-f34d420f66a3
+-----------------------+---------------------------------------------------------
| Field                 | Value
|
+-----------------------+---------------------------------------------------------
| admin_state_up        | True
```

第 2 章　OpenStack with OpenDaylight　(DevStack 編)

```
|                          |
| allowed_address_pairs    |
|                          |
| binding:vnic_type        | normal
|                          |
| device_id                |
dhcpd3377d3c-a0d1-5d71-9947-f17125c357bb-f8209267-5cf1-4073-9af5-bbbf7fe500cd
|                          |
| device_owner             | network:dhcp
|  # DHCP
| dns_assignment           | {"hostname": "host-10-11-12-2", "ip_address":
"10.11.12.2", "fqdn": "host-10-11-12-2.openstacklocal."} |
| dns_name                 |
|                          |
| extra_dhcp_opts          |
|                          |
| fixed_ips                | {"subnet_id": "8d373fbd-6134-47d1-aff6-86ee1ae7c8d1",
"ip_address": "10.11.12.2"}                             |
| id                       | 178a8b99-700f-4d2f-be35-f34d420f66a3
|                          |
| mac_address              | fa:16:3e:ef:85:07
|                          |
| name                     |
|                          |
| network_id               | f8209267-5cf1-4073-9af5-bbbf7fe500cd
|                          |
| port_security_enabled    | False
|                          |
| security_groups          |
|                          |
| status                   | ACTIVE
|                          |
| tenant_id                | 7305d84b42cb470c9b3bbb96e711cc96
|                          |
+--------------------------+------------------------------------------------
```

デバイス名と Openflow ポート番号の対応は以下の様に取得できます。DHCP サーバのポート (tap178a8b99-70) は br-int の OpenFlow ポート 1 番であることがわかります。

[コントローラノード]

```
~/devstack$ sudo ovs-vsctl --columns=name,ofport list Interface
name                : "tap178a8b99-70"   # デバイス名 (DHCP サーバ)
ofport              : 1                  # br-int の OpenFlow ポート番号

name                : br-ex
ofport              : 65534

name                : "vxlan-192.168.0.11"
ofport              : 2

name                : br-int
```

```
ofport                    : 65534
```

[コンピュートノード]

```
~/devstack$ sudo ovs-vsctl --columns=name,ofport list Interface
name                      : "vxlan-192.168.0.10"
ofport                    : 1

name                      : br-int
ofport                    : 65534
```

この状態のフローを見てみると、ノード間通信とDHCPに関するフローが登録されていることがわかります。

[コントローラノード]

```
~/devstack$ sudo ovs-ofctl dump-flows br-int -O openflow13
OFPST_FLOW reply (OF1.3) (xid=0x2):
 cookie=0x0, duration=10069.488s, table=0, n_packets=0, n_bytes=0,
tun_id=0x435,in_port=2 actions=load:0x2->NXM_NX_REG0[],goto_table:20
 cookie=0x0, duration=10069.761s, table=0, n_packets=5, n_bytes=390,
in_port=1,dl_src=fa:16:3e:ef:85:07
actions=set_field:0x435->tun_id,load:0x1->NXM_NX_REG0[],goto_table:20
 cookie=0x0, duration=10069.737s, table=0, n_packets=0, n_bytes=0,
priority=8192,in_port=1 actions=drop
 cookie=0x0, duration=12211.097s, table=0, n_packets=2015, n_bytes=227695,
dl_type=0x88cc actions=CONTROLLER:65535
 cookie=0x0, duration=12209.963s, table=0, n_packets=6, n_bytes=488, priority=0
actions=goto_table:20
 cookie=0x0, duration=12209.956s, table=20, n_packets=11, n_bytes=878, priority=0
actions=goto_table:30
 cookie=0x0, duration=12209.953s, table=30, n_packets=11, n_bytes=878, priority=0
actions=goto_table:40
 cookie=0x0, duration=10070.279s, table=40, n_packets=0, n_bytes=0,
priority=61012,udp,tp_src=68,tp_dst=67 actions=goto_table:50
 cookie=0x0, duration=12209.948s, table=40, n_packets=11, n_bytes=878, priority=0
actions=goto_table:50
 cookie=0x0, duration=12209.942s, table=50, n_packets=11, n_bytes=878, priority=0
actions=goto_table:60
 cookie=0x0, duration=12209.94s, table=60, n_packets=11, n_bytes=878, priority=0
actions=goto_table:70
 cookie=0x0, duration=12209.936s, table=70, n_packets=11, n_bytes=878, priority=0
actions=goto_table:80
 cookie=0x0, duration=12209.931s, table=80, n_packets=11, n_bytes=878, priority=0
actions=goto_table:90
 cookie=0x0, duration=12209.926s, table=90, n_packets=11, n_bytes=878, priority=0
actions=goto_table:100
 cookie=0x0, duration=12209.922s, table=100, n_packets=11, n_bytes=878, priority=0
actions=goto_table:110
 cookie=0x0, duration=10069.5s, table=110, n_packets=0, n_bytes=0,
priority=8192,tun_id=0x435 actions=drop
```

第 2 章　OpenStack with OpenDaylight（DevStack 編）

```
 cookie=0x0, duration=10069.718s, table=110, n_packets=0, n_bytes=0,
tun_id=0x435,dl_dst=fa:16:3e:ef:85:07 actions=output:1
 cookie=0x0, duration=10069.508s, table=110, n_packets=5, n_bytes=390,
priority=16383,reg0=0x1,tun_id=0x435,dl_dst=01:00:00:00:00:00/01:00:00:00:00:00
actions=output:1,output:2
 cookie=0x0, duration=10069.506s, table=110, n_packets=0, n_bytes=0,
priority=16384,reg0=0x2,tun_id=0x435,dl_dst=01:00:00:00:00:00/01:00:00:00:00:00
actions=output:1
 cookie=0x0, duration=12209.889s, table=110, n_packets=6, n_bytes=488, priority=0
actions=drop
```

［コンピュートノード］

```
~/devstack$ sudo ovs-ofctl dump-flows br-int -O OpenFlow13
OFPST_FLOW reply (OF1.3) (xid=0x2):
 cookie=0x0, duration=12869.133s, table=0, n_packets=2254, n_bytes=254702,
dl_type=0x88cc actions=CONTROLLER:65535
 cookie=0x0, duration=12867.254s, table=0, n_packets=9, n_bytes=710, priority=0
actions=goto_table:20
 cookie=0x0, duration=12867.251s, table=20, n_packets=8, n_bytes=620, priority=0
actions=goto_table:30
 cookie=0x0, duration=12867.249s, table=30, n_packets=8, n_bytes=620, priority=0
actions=goto_table:40
 cookie=0x0, duration=12867.249s, table=40, n_packets=8, n_bytes=620, priority=0
actions=goto_table:50
 cookie=0x0, duration=12867.243s, table=50, n_packets=8, n_bytes=620, priority=0
actions=goto_table:60
 cookie=0x0, duration=12867.238s, table=60, n_packets=8, n_bytes=620, priority=0
actions=goto_table:70
 cookie=0x0, duration=12867.232s, table=70, n_packets=8, n_bytes=620, priority=0
actions=goto_table:80
 cookie=0x0, duration=12867.221s, table=80, n_packets=8, n_bytes=620, priority=0
actions=goto_table:90
 cookie=0x0, duration=12867.211s, table=90, n_packets=8, n_bytes=620, priority=0
actions=goto_table:100
 cookie=0x0, duration=12867.205s, table=100, n_packets=8, n_bytes=620, priority=0
actions=goto_table:110
 cookie=0x0, duration=11271.53s, table=110, n_packets=0, n_bytes=0,
tun_id=0x435,dl_dst=fa:16:3e:ef:85:07 actions=output:1
 cookie=0x0, duration=12867.198s, table=110, n_packets=8, n_bytes=620, priority=0
actions=drop
```

最後にインスタンスを起動した時の状態を確認します。インスタンスを起動するとコンピュートノードの br-int にインスタンスのポート（tap7adff632-cc、tap2756a09c-12）が接続されます。

［コンピュートノード］

```
~/devstack$ sudo ovs-vsctl show
00790da7-431d-430c-b638-9114e2f17552
    Manager "tcp:192.168.0.10:6640"
        is_connected: true
```

2.4 仕組みの解説

```
    Bridge br-int
        Controller "tcp:192.168.0.10:6653"
            is_connected: true
        fail_mode: secure
        Port "tap7adff632-cc" # インスタンスのポート
            Interface "tap7adff632-cc"
        Port br-int
            Interface br-int
                type: internal
        Port "vxlan-192.168.0.10"
            Interface "vxlan-192.168.0.10"
                type: vxlan
                options: {key=flow, local_ip="192.168.0.11",
remote_ip="192.168.0.10"}
        Port "tap2756a09c-12" # インスタンスのポート
            Interface "tap2756a09c-12"
    ovs_version: "2.0.2"
```

2台のインスタンスのそれぞれのXML設定ファイルを見てみると、br-intのポート名（tap7adff632-cc と tap2756a09c-12）とインスタンスのデバイス名が対応していることが分かります。

[コンピュートノード]

```
~/devstack$ virsh list
 Id    Name                           State
----------------------------------------------------
 2     instance-00000001              running
 3     instance-00000002              running
~/devstack$ virsh dumpxml instance-00000001
<domain type='kvm' id='2'>
  <name>instance-00000001</name>
  <uuid>7d82ceb1-1fae-4b3c-b294-03ba5362200b</uuid>
  :
<devices>
  :
    <interface type='bridge'>
      <mac address='fa:16:3e:9e:f5:70'/>
      <source bridge='br-int'/>
      <virtualport type='openvswitch'>
        <parameters interfaceid='7adff632-ccf9-4d43-8f5d-7db454313f44'/>
      </virtualport>
      <target dev='tap7adff632-cc'/>
      <model type='virtio'/>
  :
~/devstack$ virsh dumpxml instance-00000002
<domain type='kvm' id='3'>
  <name>instance-00000002</name>
  <uuid>e6151ecd-41e1-4730-bcfa-7cbc68a86830</uuid>
  :
<devices>
```

第 2 章　OpenStack with OpenDaylight　（DevStack 編）

```
    :
  <interface type='bridge'>
      <mac address='fa:16:3e:53:88:56'/>
      <source bridge='br-int'/>
      <virtualport type='openvswitch'>
        <parameters interfaceid='2756a09c-125a-4fe8-9a26-2bde05857062'/>
      </virtualport>
      <target dev='tap2756a09c-12'/>
      <model type='virtio'/>
    :
```

　OpenFlow のポート番号を調べて、インスタンスに対して設定されたフローを確認します。詳しいフローの説明は省略しますが、インスタンスを作成すると各ノードにインスタンスの通信に関するフローが設定されることが分かると思います。

［コントローラノード］

```
~/devstack$ sudo ovs-ofctl dump-flows br-int -O openflow13
OFPST_FLOW reply (OF1.3) (xid=0x2):
 cookie=0x0, duration=17198.417s, table=0, n_packets=60, n_bytes=5204,
tun_id=0x435,in_port=2 actions=load:0x2->NXM_NX_REG0[],goto_table:20
    :
 cookie=0x0, duration=2970.026s, table=110, n_packets=15, n_bytes=1801,
tun_id=0x435,dl_dst=fa:16:3e:53:88:56 actions=output:2
 cookie=0x0, duration=2979.034s, table=110, n_packets=15, n_bytes=1801,
tun_id=0x435,dl_dst=fa:16:3e:9e:f5:70 actions=output:2
 cookie=0x0, duration=17198.647s, table=110, n_packets=24, n_bytes=2084,
tun_id=0x435,dl_dst=fa:16:3e:ef:85:07 actions=output:1
    :
```

［コンピュートノード］

```
~/devstack$ sudo ovs-vsctl --columns=name,ofport list Interface
name                : "tap2756a09c-12"   # vm02
ofport              : 3

name                : br-int
ofport              : 65534

name                : "vxlan-192.168.11.43"
ofport              : 1

name                : "tap7adff632-cc"   # vm01
ofport              : 2

~/devstack$ sudo ovs-ofctl dump-flows br-int -O OpenFlow13
OFPST_FLOW reply (OF1.3) (xid=0x2):
 cookie=0x0, duration=3110.533s, table=0, n_packets=30, n_bytes=3602,
tun_id=0x435,in_port=1 actions=load:0x2->NXM_NX_REG0[],goto_table:20
 cookie=0x0, duration=3110.564s, table=0, n_packets=30, n_bytes=2602,
in_port=2,dl_src=fa:16:3e:9e:f5:70
```

2.4 仕組みの解説

```
actions=set_field:0x435->tun_id,load:0x1->NXM_NX_REG0[],goto_table:20
 cookie=0x0, duration=3101.583s, table=0, n_packets=30, n_bytes=2602,
in_port=3,dl_src=fa:16:3e:53:88:56
actions=set_field:0x435->tun_id,load:0x1->NXM_NX_REG0[],goto_table:20
 cookie=0x0, duration=3101.577s, table=0, n_packets=0, n_bytes=0,
priority=8192,in_port=3 actions=drop
 cookie=0x0, duration=3110.559s, table=0, n_packets=0, n_bytes=0,
priority=8192,in_port=2 actions=drop
 cookie=0x0, duration=18927.461s, table=0, n_packets=3466, n_bytes=391658,
dl_type=0x88cc actions=CONTROLLER:65535
 cookie=0x0, duration=18925.582s, table=0, n_packets=9, n_bytes=710, priority=0
actions=goto_table:20
 cookie=0x0, duration=18925.579s, table=20, n_packets=98, n_bytes=9426, priority=0
actions=goto_table:30
 cookie=0x0, duration=18925.577s, table=30, n_packets=98, n_bytes=9426, priority=0
actions=goto_table:40
 cookie=0x0, duration=3110.566s, table=40, n_packets=13, n_bytes=1672,
priority=61007,ip,dl_src=fa:16:3e:9e:f5:70 actions=goto_table:50
 cookie=0x0, duration=3101.594s, table=40, n_packets=13, n_bytes=1672,
priority=61007,ip,dl_src=fa:16:3e:53:88:56 actions=goto_table:50
 cookie=0x0, duration=3110.595s, table=40, n_packets=0, n_bytes=0,
priority=36001,ip,in_port=2,dl_src=fa:16:3e:9e:f5:70,nw_src=10.11.12.3
actions=goto_table:50
 cookie=0x0, duration=3101.61s, table=40, n_packets=0, n_bytes=0,
priority=36001,ip,in_port=3,dl_src=fa:16:3e:53:88:56,nw_src=10.11.12.4
actions=goto_table:50
 cookie=0x0, duration=18925.577s, table=40, n_packets=72, n_bytes=6082, priority=0
actions=goto_table:50
 cookie=0x0, duration=3110.62s, table=40, n_packets=0, n_bytes=0,
priority=61011,udp,in_port=2,tp_src=67,tp_dst=68 actions=drop
 cookie=0x0, duration=3101.619s, table=40, n_packets=0, n_bytes=0,
priority=61011,udp,in_port=3,tp_src=67,tp_dst=68 actions=drop
 cookie=0x0, duration=18925.571s, table=50, n_packets=98, n_bytes=9426, priority=0
actions=goto_table:60
 cookie=0x0, duration=18925.566s, table=60, n_packets=98, n_bytes=9426, priority=0
actions=goto_table:70
 cookie=0x0, duration=18925.56s, table=70, n_packets=98, n_bytes=9426, priority=0
actions=goto_table:80
 cookie=0x0, duration=18925.549s, table=80, n_packets=98, n_bytes=9426, priority=0
actions=goto_table:90
 cookie=0x0, duration=3101.6s, table=90, n_packets=13, n_bytes=1717,
priority=61007,ip,dl_dst=fa:16:3e:53:88:56 actions=goto_table:100
 cookie=0x0, duration=3110.595s, table=90, n_packets=13, n_bytes=1717,
priority=61007,ip,dl_dst=fa:16:3e:9e:f5:70 actions=goto_table:100
 cookie=0x0, duration=18925.539s, table=90, n_packets=72, n_bytes=5992, priority=0
actions=goto_table:100
 cookie=0x0, duration=3110.679s, table=90, n_packets=0, n_bytes=0,
priority=61006,udp,dl_src=fa:16:3e:ef:85:07,tp_src=67,tp_dst=68
actions=goto_table:100
 cookie=0x0, duration=18925.533s, table=100, n_packets=98, n_bytes=9426, priority=0
actions=goto_table:110
 cookie=0x0, duration=3110.537s, table=110, n_packets=0, n_bytes=0,
priority=8192,tun_id=0x435 actions=drop
```

```
 cookie=0x0, duration=3101.572s, table=110, n_packets=15, n_bytes=1801,
tun_id=0x435,dl_dst=fa:16:3e:53:88:56 actions=output:3
 cookie=0x0, duration=3110.557s, table=110, n_packets=15, n_bytes=1801,
tun_id=0x435,dl_dst=fa:16:3e:9e:f5:70 actions=output:2
 cookie=0x0, duration=17329.858s, table=110, n_packets=24, n_bytes=2084,
tun_id=0x435,dl_dst=fa:16:3e:ef:85:07 actions=output:1
 cookie=0x0, duration=3110.545s, table=110, n_packets=36, n_bytes=3120,
priority=16383,reg0=0x1,tun_id=0x435,dl_dst=01:00:00:00:00:00/01:00:00:00:00:00
actions=output:2,output:1,output:3
 cookie=0x0, duration=3110.549s, table=110, n_packets=0, n_bytes=0,
priority=16384,reg0=0x2,tun_id=0x435,dl_dst=01:00:00:00:00:00/01:00:00:00:00:00
actions=output:2,output:3
 cookie=0x0, duration=18925.526s, table=110, n_packets=8, n_bytes=620, priority=0
actions=drop
```

ODL と OpenStack の連携の仕組みに関する説明は以上です。

2.5 SSHでインスタンスに接続したい

以降の説明は構築やデバッグのためのノウハウを紹介します。

ホストマシンからインスタンスに SSH で接続したい場合、外部ネットワークを作ってインスタンスに Floating IP を割り振る必要があります。先ほど作成したインスタンス「vm01」に Floating IP を割り振って SSH で接続してみます。Cirros イメージのユーザ名とパスワードはそれぞれ「cirros」と「cubswin:)」となっています。

```
~/devstack$ source openrc admin admin
~/devstack$ neutron net-create ext-net --router:external       # 外部ネットワークを作成
する
~/devstack$ neutron subnet-create ext-net 172.16.1.0/24        # サブネットを作成する
~/devstack$ sudo ip addr add 172.16.1.1/24 dev br-ex           # br-ex へ IP アドレスを
割り当てる
~/devstack$ source openrc demo demo
~/devstack$ neutron router-create router1                      # 仮想ルータを作成する
~/devstack$ neutron router-interface-add router1 net01-subnet  # 仮想ルータへ
net01-subnet のインターフェースを追加する
~/devstack$ neutron router-gateway-set router1 ext-net         # 仮想ルータへゲートウェ
イを設定する
~/devstack$ neutron floatingip-create ext-net                  # フローティング IP を作
成する
Created a new floatingip:
+---------------------+--------------------------------------+
| Field               | Value                                |
+---------------------+--------------------------------------+
| fixed_ip_address    |                                      |
| floating_ip_address | 172.16.1.4                           |
| floating_network_id | 196a6d94-a43b-4d77-bd99-2fa0e8ab311b |
```

```
| id                | ac605256-6aae-4391-a52c-05dc5aac1a8b |
| port_id           |                                      |
| router_id         |                                      |
| status            | DOWN                                 |
| tenant_id         | 7305d84b42cb470c9b3bbb96e711cc96     |
+-------------------+--------------------------------------+
~/devstack$ neutron port-list
+--------------------------------------+------+-------------------+------------------
| id                                   | name | mac_address       | fixed_ips
|
| 178a8b99-700f-4d2f-be35-f34d420f66a3 |      | fa:16:3e:ef:85:07 | {"subnet_id": "8d373fbd-6134-47d1-aff6-86ee1ae7c8d1", "ip_address": "10.11.12.2"} |
| 2756a09c-125a-4fe8-9a26-2bde05857062 |      | fa:16:3e:53:88:56 | {"subnet_id": "8d373fbd-6134-47d1-aff6-86ee1ae7c8d1", "ip_address": "10.11.12.4"} |
| 7adff632-ccf9-4d43-8f5d-7db454313f44 |      | fa:16:3e:9e:f5:70 | {"subnet_id": "8d373fbd-6134-47d1-aff6-86ee1ae7c8d1", "ip_address": "10.11.12.3"} |
| f72c630e-7389-4b91-bbda-5c27e768ad7b |      | fa:16:3e:bb:4f:57 | {"subnet_id": "8d373fbd-6134-47d1-aff6-86ee1ae7c8d1", "ip_address": "10.11.12.1"} |
+--------------------------------------+------+-------------------+------------------
~/devstack$ neutron floatingip-associate  ac605256-6aae-4391-a52c-05dc5aac1a8b
7adff632-ccf9-4d43-8f5d-7db454313f44   # vm01 のポートへフローティング IP を割り当てる
~/devstack$ ssh cirros@172.16.1.4
The authenticity of host '172.16.1.4 (172.16.1.4)' can't be established.
RSA key fingerprint is ef:bc:ba:b7:21:ca:3b:45:80:1c:12:e3:c6:8c:96:5e.
Are you sure you want to continue connecting (yes/no)? yes
Warning: Permanently added '172.16.1.4' (RSA) to the list of known hosts.
cirros@172.16.1.4's password:   # パスワード：cubswin:) を入力する (表示はされない)
$ hostname
vm01
```

2.6 ログの出力先

ODL の karaf ログはコントローラノードの/opt/stack/logs/screen-karaf.txt に出力されます。OpenStack の各コンポーネントのログも/opt/stack/logs/ディレクトリに出力されるので必要であれば参照して下さい。

2.7 （おまけ）シングルノードでの構築

今回はコントローラノードとコンピュートノードの 2 台の仮想マシンを使ってマルチノードで構築しましたが、シングルノードで構築してみたいという方は以下の URL の local.conf を使って挑戦してみてください。

https://github.com/YujiAzama/opendaylight-openstack-integration/tree/master/allinone

/single-node/local.conf

　本章は OpenDaylight と OpenStack を連携させるために、もっとも簡単な DevStack を使った構築を行いました。次章は手作業で OpenDaylight をインストールし OpenStack と連携させる手順をご紹介します。

第3章 OpenStack with OpenDaylight（手動構築編）

　前章は DevStack を使って OpenDaylight（以降、ODL と省略）や OpenStack の構築を行った上で、連携の仕組みや簡単な使い方について触れました。DevStack という自動構築ツールを使ったので簡単だったと思います。そこで本章は、DevStack による構築は OpenStack にとどめ、ODL の手動構築にチャレンジしてみます。

3.1 環境・バージョンについて

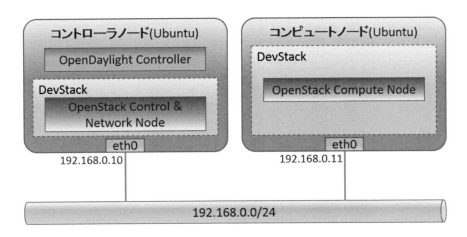

図 3.1　ノード構成図

　ノード構成やネットワーク構成などは前回と同じ環境ですが、ODL の構築は DevStack の管理

外となります。ODL のバージョンは前回は DevStack のデフォルトである「0.3.5-SNAPSHOT」を使用しましたが、今回は Lithium のリリースである版「0.3.4-Lithium-SR4」を使うことにします。ODL と OpenStack 連携の動作に違いはありません。

3.2 構築してみよう

全ノード共通の設定

それでは早速構築を始めていきます。先に、全ノードで共通の設定を実施しておきます。

構築を簡単にするために、ファイアーウォールは無効化しておきます。

```
~$ sudo ufw disable
```

次に DevStack をダウンロードします。その際、git コマンドが必要なので入っていなければ必要に応じてインストールしてください。ブランチ名は stable/liberty を指定します。

```
~$ sudo apt-get install git
~$ git clone --branch stable/liberty https://github.com/openstack-dev/devstack.git
~$ cd devstack/
```

続いて、コントローラノードの構築に入ります。

コントローラノードの構築

JDK のインストール

ODL は Java 言語によって実装されています。そのため Java の実行環境が必要になりますので JDK(Java Development Kit) をインストールします。今回インストールする Lithium では JDK 7 が必要です。ちなみに、2016 年 2 月にリリースされたバージョンである Beryllium では JDK 8 が必要になります。

```
~$ sudo apt-get install openjdk-7-jre
```

OpenDaylight のダウンロード

ODL のパッケージは Sonatype NEXUS パッケージリポジトリマネジメントソフトウェアで管理されています。そのリポジトリから Lithium リリースのパッケージをダウンロードし解凍します。

3.2 構築してみよう

```
~$ wget
https://nexus.opendaylight.org/content/repositories/opendaylight.release/org/openday
light/integration/distribution-karaf/0.3.4-Lithium-SR4/distribution-karaf-0.3.4-
Lithium-SR4.zip
~$ unzip -u -o distribution-karaf-0.3.4-Lithium-SR4.zip
~$ cd distribution-karaf-0.3.4-Lithium-SR4/
```

OpenDaylight を起動する

ODL を起動するには、以下のように start スクリプトを実行するだけです。

```
~/distribution-karaf-0.3.4-Lithium-SR4$ ./bin/start   # OpenDaylight を起動する
```

必要な機能をインストールする

ODL と OpenStack を連携させるには必要なプラグインをインストールしなければなりません。プラグインのインストールは client スクリプトを使って OpenDaylight Karaf Shell にログインして行います。以下の様に、odl-ovsdb-openstack とその他の依存関係にあるパッケージをインストールします。

```
~/distribution-karaf-0.3.4-Lithium-SR4$ ./bin/client -u karaf   # OpenDaylight Karaf Shell にログイン
client: JAVA_HOME not set; results may vary
Logging in as karaf
308 [sshd-SshClient[1c3a0533]-nio2-thread-2] WARN
org.apache.sshd.client.keyverifier.AcceptAllServerKeyVerifier - Server at
[/0.0.0.0:8101, DSA, f9:9d:15:00:e5:c9:4f:3f:a4:6f:d3:65:02:3c:b5:e3] presented
unverified {} key: {}

        _____                       _____                      .__  .__       .__     __
        \_____  \ _____   ____   ____ _____ \ _____  ___.__.|  | |__| ____ |  |___/  |_
         /   |   \\____ \_/ __ \ /    \ |    |  \\__  \<   |  ||  | |  |/ ___\|  |  \   __\
        /    |    \  |_> >  ___/|   |  \|    `   \/ __ \\___  ||  |_|  / /_/  >   Y  \  |
        _____  /   __/ \___  >___|  /_____  (____  / ____||____/__\___  /|___|  /__|
                \/|__|        \/     \/        \/     \/\/            /_____/      \/

Hit '<tab>' for a list of available commands
and '[cmd] --help' for help on a specific command.
Hit '<ctrl-d>' or type 'system:shutdown' or 'logout' to shutdown OpenDaylight.

opendaylight-user@root>feature:install odl-restconf-all odl-aaa-authn odl-dlux-core
odl-mdsal-apidocs odl-ovsdb-openstack   # 機能をインストールする
opendaylight-user@root>logout
```

以上で、ODL の構築は完了です。

ちなみに、実行中の ODL を停止させる場合には stop スクリプトを実行します。

```
~/distribution-karaf-0.3.4-Lithium-SR4$ ./bin/stop    # OpenDaylight を停止する
```

OpenStack コントローラノードの構築

続いて、OpenStack の構築を行います。以下の URL から local.conf のサンプルをダウンロードしてください。

https://github.com/YujiAzama/opendaylight-openstack-integration/blob/master/externalodl/multi-node/control/local.conf

ダウンロードした local.conf をテキストエディタで開くと以下の様になっています。まず、お使いの環境に合わせて HOST_IP をコントローラノードの IP アドレスに変更してください。

```
~/devstack$ vim local.conf
   :
# IP Details
HOST_IP=192.168.0.10
SERVICE_HOST=$HOST_IP
   :
# Neutron
   :
enable_plugin networking-odl http://git.openstack.org/openstack/networking-odl ${BRANCH_NAME}

# OpenDaylight Details
ODL_MODE=externalodl
ODL_PORT=8181

[[post-config|/etc/neutron/plugins/ml2/ml2_conf.ini]]
[ml2_odl]
password=admin
username=admin
url="http://${ODL_MGR_IP}:${ODL_PORT}/controller/nb/v2/neutron"
```

今回の DevStack での構築には、ODL に関する設定のポイントが2つあります。1つ目は動作モードを externalodl とすることです。今回、ODL は手動で構築しましたので DevStack の管理外とする必要があります。この場合は externalodl を指定します。2つ目は ml2_conf.ini にユーザ名やパスワード、エンドポイントの URL などの ODL の認証情報を追記することです。これらは DevStack が各ノードのブリッジの設定などをするための API リクエストに必要な情報となります。ユーザ名とパスワードに指定している「admin」は ODL のデフォルト値なので ODL の構築時に変更していなければ、こちらも変更しないで下さい。

後は、stack.sh スクリプトを実行するだけです。以下の様なメッセージが出力されれば終了です。

```
~/devstack$ ./stack.sh
  :
This is your host IP address: 192.168.0.10
This is your host IPv6 address: ::1
Horizon is now available at http://192.168.0.10/dashboard
Keystone is serving at http://192.168.0.10:5000/
The default users are: admin and demo
The password: password
```

　以上でコントローラノードの構築は完了です。

コンピュートノードの構築

　コンピュートノードの構築は前回の記事とまったく同じになります。下記 URL より local.conf のサンプルをダウンロードします。

　https://github.com/YujiAzama/opendaylight-openstack-integration/blob/master/externalodl/multi-node/compute/local.conf

　ダウンロードした local.conf を開き、HOST_IP をコンピュートノードの IP アドレスへ、SERVICE_HOST をコントローラノードの IP アドレスへ変更した後、stack.sh スクリプトを実行します。以下の様なメッセージが出力されれば終了です。

```
~/devstack$ ./stack.sh
  :
This is your host IP address: 192.168.0.11
This is your host IPv6 address: ::1
2016-01-13 07:21:51.305 | stack.sh completed in 56 seconds.
```

　以上で、ODL を手動で構築する際の手順は全て終了です。

　ODL のダウンロードと起動・停止、プラグインのインストールさえできれば、手動構築の難易度もそれほど高くないことを理解していただけましたでしょうか。ぜひこれを参考に OpenDaylight を使い倒してみてください。

3.3　tcpdumpを使ってデバッグするポイント

　パケットが正しく流れているかなどを調査する際、tcpdump コマンドを使う事が多いと思います。ODL と OpenStack 連携において、VM の通信がうまくいっていない場合には以下のポイントで tcpdump を取得すると良いでしょう。

第 3 章　OpenStack with OpenDaylight（手動構築編）

- (1) インスタンスの tap が接続されている br-int のポート
  - インスタンスのポートまでパケットが届いているか確認します。
- (2), (3) コンピュートノードとコンピュートノードの eth0
  - コントローラノードとコンピュートノードの NIC がパケットを正しく送受信しているか確認します。ここで問題が発生していると VxLAN も正しくトンネリングされません。
- (4) qdhcp の tap が接続されている br-int のポート
  - DHCP が正しく機能しているか確認します。インスタンスからのリクエストを受けているか、DHCP サーバはレスポンスを返しているかを確認して下さい。

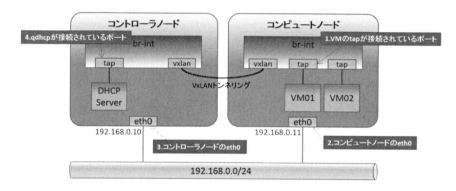

図 3.2　tcpdump

例として、コンピュートノードでインスタンスの tap が接続されている br-int のポート見る場合、以下の様に実行します。

[コンピュートノード]

```
~$ sudo ovs-vsctl show
00790da7-431d-430c-b638-9114e2f17552
    Manager "tcp:192.168.11.43:6640"
        is_connected: true
    Bridge br-int
        Controller "tcp:192.168.11.43:6653"
            is_connected: true
        fail_mode: secure
        Port br-int
            Interface br-int
                type: internal
        Port "vxlan-192.168.11.43"
            Interface "vxlan-192.168.11.43"
```

```
                type: vxlan
                options: {key=flow, local_ip="192.168.11.44",
remote_ip="192.168.11.43"}
        Port "tapeb836a62-e6"
            Interface "tapeb836a62-e6"   # tap が接続している Interface
    ovs_version: "2.0.2"
~$
~$ sudo tcpdump -n -i tapeb836a62-e6   # tap が接続している Interface を指定
tcpdump: WARNING: tapeb836a62-e6: no IPv4 address assigned
tcpdump: verbose output suppressed, use -v or -vv for full protocol decode
listening on tapeb836a62-e6, link-type EN10MB (Ethernet), capture size 65535 bytes
15:22:01.870876 LLDP, length 97: openflow:11918597943880
15:22:06.871614 LLDP, length 97: openflow:11918597943880
    :
```

3.4　(おまけ) シングルノードでの構築

　今回はコントローラノードとコンピュートノードの 2 台の仮想マシンを使ってマルチノードで構築しましたが、シングルノードで構築してみたいという方は以下の URL の local.conf をベースにすると良いでしょう。

　https://github.com/YujiAzama/opendaylight-openstack-integration/tree/master/externalodl/single-node/local.conf

　本章は、OpenDaylight の手動構築に挑戦しました。次章は ODL の RESTCONF と呼ばれる Northbound API について触れてみたいと思います。

第4章 RESTCONF APIを使ってフローを書き換えてみる

　OpenDaylightの特徴の一つとして、多くのNorthbound Inrerface（NBI）が用意されている点があげられます。MD-SALがインターフェースを抽象化することで、OpenDaylightを利用するさまざまなアプリケーションからのAPI呼び出しを容易にできるようになっています。第2回と第3回ではOpenDaylightとOpenStackを連携させた環境の構築を行いましたが、OpenStackにおいてもNeutronプラグイン経由でOpenDaylightに用意されているAPIを呼び出すことで、ネットワークに関する様々なコントロールができるようになっています。

　そこで、本章はOpenDaylightのRESTCONF APIに注目します。RESTCONF APIを利用して、OpenDaylightのデータストアに登録されているフローの取得と更新を行ってみます。

4.1 RESTCONF API

　そもそもRESTCONF APIとは、NETCONFデータストア内のYANGで定義されたデータにアクセスするためのHTTPを使ったRESTライクなプロトコルのことです。OpenDaylightでもRESTCONF APIを使えば、OpenDaylightのデータストアに登録されたフローをコントロールできます。

4.2 OpenDaylight RESTCONF API Documentation

　OpenDaylightを起動すると、RESTCONF API ExplorerがOSGi bundleとして実行されるようになっています。RESTCONF API ExplorerはAPIドキュメントなのですが、Swagger

の仕様に基づいており、レンダリングされて Swagger UI として表示されます。OpenDaylight を実行しているノードに対して、ブラウザから以下の様にアクセスすると、API の一覧を表示したり、テストリクエストを投げたりすることができます。

http://192.168.0.10:8181/apidoc/explorer/index.html

図 4.1　RESTCONF API Explorer

4.3　RESTCONF APIを使ってみる

それでは、RESTCONF API を試してみます。環境は第 2 章で構築した OpenDaylight +OpenStack 環境を使います。

フローの取得

RESTCONF の HTTP リクエストとレスポンスの内容は、非常に長くなる場合があり記述するのが面倒です。そこで、ここではクライアントコマンドでオペレーションしていきます。まず、python-odlclient をインストールします。

```
~$ sudo pip install python-odlclient
```

フローを取得するためには、取得したいノードのノード ID を知る必要があります。そこで、まず最初にノード一覧を取得してコンピュートノードのノード ID を確認します。

4.3 RESTCONF API を使ってみる

```
~$ export ODL_HOST=192.168.0.10
~$ odl node list
+--------------------------+--------------+-----------------+-------------+---------+
| id                       | ip_address   | connector_count | table_count | hardware
| software |
+--------------------------+--------------+-----------------+-------------+---------+
| openflow:267140174830659 | 192.168.0.11 | 4               | 11          | Open
vSwitch | 2.0.2     | # コンピュートノード
| openflow:235547676672585 | 192.168.0.10 | 5               | 11          | Open
vSwitch | 2.0.2     | # コントローラノード
+--------------------------+--------------+-----------------+-------------+---------+
```

ちなみに、ノード一覧を取得するために curl コマンドを使ってリクエストすると、以下の様に膨大な量のレスポンスが返ってくるため、うまく加工しない限り人間が読めたものではありません。

```
~$ curl -s -H "Accept: application/json" -u "admin:admin" -X GET
http://192.168.0.10:8181/restconf/operational/opendaylight-inventory:nodes
{"nodes":{"node":[{"id":"openflow:267140174830659","node-connector":[{"id":"openflow:
267140174830659:1","flow-node-inventory:hardware-address":"76:6B:A2:F5:73:75",……
..}
```

ノードごとのノード ID がわかったので、コンピュートノードのフローを取得してみます。

```
~/devstack$ odl flow list openflow:267140174830659
+-------------------------------------------------+----------+----------+---------+
| id                                              | table_id | priority | match
| instructions
|
+-------------------------------------------------+----------+----------+---------+
| LLDP                                            | 0        | 32768    |
ethernet-match: {"ethernet-type": {"type": 35020}}
| output-action: {"max-length": 65535, "output-node-connector": "CONTROLLER"}
|
| LocalMac_1056_7_fa:16:3e:db:c3:fd               | 0        | 32768    | in-port:
"openflow:267140174830659:7", ethernet-match: {"ethernet-source": {"address":
"FA:16:3E:DB:C3:FD"}}
| set-field: {"tunnel": {"tunnel-id": 1056}},
openflowplugin-extension-nicira-action:nx-reg-load: {"dst": {"start": 0, "end": 31,
"nx-reg": "nicira-match:nxm-nx-reg0"}, "value": 1}, go-to-table: {"table_id": 20} |
| LocalMac_1056_6_fa:16:3e:c5:8b:b9               | 0        | 32768    | in-port:
"openflow:267140174830659:6", ethernet-match: {"ethernet-source": {"address":
"FA:16:3E:C5:8B:B9"}}
| set-field: {"tunnel": {"tunnel-id": 1056}},
openflowplugin-extension-nicira-action:nx-reg-load: {"dst": {"start": 0, "end": 31,
"nx-reg": "nicira-match:nxm-nx-reg0"}, "value": 1}, go-to-table: {"table_id": 20} |
| TunnelIn_1056_1                                 | 0        | 32768    | tunnel:
{"tunnel-id": 1056}, in-port: "openflow:267140174830659:1"
| openflowplugin-extension-nicira-action:nx-reg-load: {"dst": {"start": 0, "end":
```

第 4 章　RESTCONF API を使ってフローを書き換えてみる

```
31, "nx-reg": "nicira-match:nxm-nx-reg0"}, "value": 2}, go-to-table: {"table_id":
20}                                                    |
| DropFilter_7                                         | 0        | 8192     | in-port:
"openflow:267140174830659:7"
| Drop
|
| DropFilter_6                                         | 0        | 8192     | in-port:
"openflow:267140174830659:6"
| Drop
|
| DEFAULT_PIPELINE_FLOW_0                              | 0        | 0        |
| go-to-table: {"table_id": 20}
|
| DEFAULT_PIPELINE_FLOW_20                             | 20       | 0        |
| go-to-table: {"table_id": 30}
|
| DEFAULT_PIPELINE_FLOW_30                             | 30       | 0        |
| go-to-table: {"table_id": 40}
|
| Egress_DHCP_Client_Permit_                           | 40       | 61012    |
udp-destination-port: 67, ip-match: {"ip-protocol": 17}, udp-source-port: 68,
ethernet-match: {"ethernet-type": {"type": 2048}}
| go-to-table: {"table_id": 50}
|
| Egress_DHCP_Server_7_DROP_                           | 40       | 61011    | in-port:
"openflow:267140174830659:7", udp-destination-port: 68, ethernet-match:
{"ethernet-type": {"type": 2048}}, udp-source-port: 67, ip-match: {"ip-protocol":
17}
| Drop
|
| Egress_DHCP_Server_6_DROP_                           | 40       | 61011    | in-port:
"openflow:267140174830659:6", udp-destination-port: 68, ethernet-match:
{"ethernet-type": {"type": 2048}}, udp-source-port: 67, ip-match: {"ip-protocol":
17}
| Drop
|
| #UF$TABLE*40-9                                       | 40       | 61010    |
arp-source-hardware-address: {"address": "FA:16:3E:C5:8B:B9"}, ethernet-match:
{"ethernet-type": {"type": 2054}}
| go-to-table: {"table_id": 50}
|
| #UF$TABLE*40-11                                      | 40       | 61010    |
arp-source-hardware-address: {"address": "FA:16:3E:DB:C3:FD"}, ethernet-match:
{"ethernet-type": {"type": 2054}}
| go-to-table: {"table_id": 50}
|
| Egress_IP1056_fa:16:3e:c5:8b:b9_Permit_              | 40       | 61007    |
ethernet-match: {"ethernet-source": {"address": "FA:16:3E:C5:8B:B9"},
"ethernet-type": {"type": 2048}}
| go-to-table: {"table_id": 50}
|
| Egress_IP1056_fa:16:3e:db:c3:fd_Permit_              | 40       | 61007    |
ethernet-match: {"ethernet-source": {"address": "FA:16:3E:DB:C3:FD"},
```

```
"ethernet-type": {"type": 2048}}
| go-to-table: {"table_id": 50}
|
| Egress_Allow_VM_IP_MAC_6fa:16:3e:c5:8b:b9_Permit_  | 40        | 36001     | in-port:
"openflow:267140174830659:6", ipv4-source: "10.11.12.7/32", ethernet-match:
{"ethernet-source": {"address": "FA:16:3E:C5:8B:B9"}, "ethernet-type": {"type":
2048}}
| go-to-table: {"table_id": 50}
|
| Egress_Allow_VM_IP_MAC_7fa:16:3e:db:c3:fd_Permit_  | 40        | 36001     | in-port:
"openflow:267140174830659:7", ipv4-source: "10.11.12.8/32", ethernet-match:
{"ethernet-source": {"address": "FA:16:3E:DB:C3:FD"}, "ethernet-type": {"type":
2048}}
| go-to-table: {"table_id": 50}
|
| DEFAULT_PIPELINE_FLOW_40                           | 40        | 0         |
| go-to-table: {"table_id": 50}
|
| DEFAULT_PIPELINE_FLOW_50                           | 50        | 0         |
| go-to-table: {"table_id": 60}
|
| DEFAULT_PIPELINE_FLOW_60                           | 60        | 0         |
| go-to-table: {"table_id": 70}
|
| DEFAULT_PIPELINE_FLOW_70                           | 70        | 0         |
| go-to-table: {"table_id": 80}
|
| DEFAULT_PIPELINE_FLOW_80                           | 80        | 0         |
| go-to-table: {"table_id": 90}
|
| #UF$TABLE*90-10                                    | 90        | 61010     |
arp-target-hardware-address: {"address": "FA:16:3E:C5:8B:B9"}, ethernet-match:
{"ethernet-type": {"type": 2054}}
| go-to-table: {"table_id": 100}
|
| #UF$TABLE*90-12                                    | 90        | 61010     |
arp-target-hardware-address: {"address": "FA:16:3E:DB:C3:FD"}, ethernet-match:
{"ethernet-type": {"type": 2054}}
| go-to-table: {"table_id": 100}
|
| Ingress_IP1056_fa:16:3e:c5:8b:b9_Permit_           | 90        | 61007     |
ethernet-match: {"ethernet-type": {"type": 2048}, "ethernet-destination":
{"address": "FA:16:3E:C5:8B:B9"}}
| go-to-table: {"table_id": 100}
|
| Ingress_IP1056_fa:16:3e:db:c3:fd_Permit_           | 90        | 61007     |
ethernet-match: {"ethernet-type": {"type": 2048}, "ethernet-destination":
{"address": "FA:16:3E:DB:C3:FD"}}
| go-to-table: {"table_id": 100}
|
| Ingress_DHCP_Server1056_fa:16:3e:57:9e:07_Permit_  | 90        | 61006     |
udp-destination-port: 68, ip-match: {"ip-protocol": 17}, udp-source-port: 67,
ethernet-match: {"ethernet-source": {"address": "FA:16:3E:57:9E:07"},
```

第4章　RESTCONF API を使ってフローを書き換えてみる

```
 "ethernet-type": {"type": 2048}}
| go-to-table: {"table_id": 100}
|
| DEFAULT_PIPELINE_FLOW_90                            | 90        | 0         |
| go-to-table: {"table_id": 100}
|
| DEFAULT_PIPELINE_FLOW_100                           | 100       | 0         |
| go-to-table: {"table_id": 110}
|
| UcastOut_1056_7_fa:16:3e:db:c3:fd                   | 110       | 32768     | tunnel:
{"tunnel-id": 1056}, ethernet-match: {"ethernet-destination": {"address":
"FA:16:3E:DB:C3:FD"}}
| output-action: {"max-length": 0, "output-node-connector": "7"}
|
| UcastOut_1056_6_fa:16:3e:c5:8b:b9                   | 110       | 32768     | tunnel:
{"tunnel-id": 1056}, ethernet-match: {"ethernet-destination": {"address":
"FA:16:3E:C5:8B:B9"}}
| output-action: {"max-length": 0, "output-node-connector": "6"}
|
| TunnelOut_1056_1_fa:16:3e:57:9e:07                  | 110       | 32768     | tunnel:
{"tunnel-id": 1056}, ethernet-match: {"ethernet-destination": {"address":
"FA:16:3E:57:9E:07"}}
| output-action: {"max-length": 0, "output-node-connector": "1"}
|
| TunnelOut_1056_1_fa:16:3e:da:ac:59                  | 110       | 32768     | tunnel:
{"tunnel-id": 1056}, ethernet-match: {"ethernet-destination": {"address":
"FA:16:3E:DA:AC:59"}}
| output-action: {"max-length": 0, "output-node-connector": "1"}
|
| BcastOut_1056                                       | 110       | 16384     | tunnel:
{"tunnel-id": 1056}, ethernet-match: {"ethernet-destination": {"mask":
"01:00:00:00:00:00", "address": "01:00:00:00:00:00"}},
openflowplugin-extension-general:extension-list: [{"extension-key":
"openflowplugin-extension-nicira-match:nxm-nx-reg0-key", "extension":
{"openflowplugin-extension-nicira-match:nxm-nx-reg": {"reg":
"nicira-match:nxm-nx-reg0", "value": 2}}}] | output-action: {"max-length": 0,
"output-node-connector": "6"}, output-action: {"max-length": 0,
"output-node-connector": "7"}
|
| TunnelFloodOut_1056                                 | 110       | 16383     | tunnel:
{"tunnel-id": 1056}, ethernet-match: {"ethernet-destination": {"mask":
"01:00:00:00:00:00", "address": "01:00:00:00:00:00"}},
openflowplugin-extension-general:extension-list: [{"extension-key":
"openflowplugin-extension-nicira-match:nxm-nx-reg0-key", "extension":
{"openflowplugin-extension-nicira-match:nxm-nx-reg": {"reg":
"nicira-match:nxm-nx-reg0", "value": 1}}}] | output-action: {"max-length": 0,
"output-node-connector": "6"}, output-action: {"max-length": 0,
"output-node-connector": "1"}, output-action: {"max-length": 0,
"output-node-connector": "7"}                        |
| LocalTableMiss_1056                                 | 110       | 8192      | tunnel:
{"tunnel-id": 1056}
| Drop
|
```

```
| DEFAULT_PIPELINE_FLOW_110                              | 110      | 0        |
| Drop                                                   |          |          |
|                                                        |          |          |
+--------------------------------------------------------+----------+----------+---------
```

フローの変更

今回は 2 台のインスタンス（vm01、vm-mirror）を使って、パケットのミラーリングをRESTCONF API から行ってみます。インスタンス vm01 宛のパケットをインスタンスvm-mirror にミラーリングするためフローの設定をします。

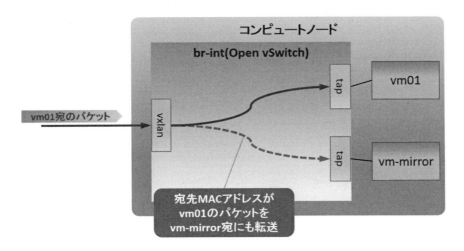

図 4.2　パケットミラーリング

まず、外部ネットワークとテナントネットワークを作成します。

```
~/devstack$ source openrc admin admin
~/devstack$ neutron net-create ext-net --router:external
~/devstack$ neutron subnet-create ext-net 172.16.1.0/24
~/devstack$ sudo ip addr add 172.16.1.1/24 dev br-ex
~/devstack$ source openrc demo demo
~/devstack$ neutron net-create net01
~/devstack$ neutron subnet-create net01 10.11.12.0/24 --name net01-subnet
~/devstack$ neutron router-create router1
~/devstack$ neutron router-interface-add router1 net01-subnet
~/devstack$ neutron router-gateway-set router1 ext-net
```

ネットワーク上のパケットをキャプチャーするツールとして tcpdump をよく利用すると思います。今回も tcpdump を利用しますが、現在 Glance イメージとして登録されている cirros は

第 4 章　RESTCONF API を使ってフローを書き換えてみる

非常にミニマムなクラウド OS であるため tcpdump コマンドがインストールされていません。そこで、Ubuntu のイメージを新規に登録することにします。

```
~/devstack$ wget
https://cloud-images.ubuntu.com/trusty/current/trusty-server-cloudimg-amd64-disk1.img
# Ubuntu のイメージをダウンロード
~/devstack$ source openrc admin admin
~/devstack$ glance image-create --name ubuntu --disk-format=qcow2
--container-format=bare \
--visibility public --file ./trusty-server-cloudimg-amd64-disk1.img
# Ubuntu イメージの登録
```

Ubuntu のインスタンスを起動し、Floating IP を割り当てます。

```
~/devstack$ source openrc demo demo
~/devstack$ nova keypair-add mykey > mykey                                    # インス
タンスへの SSH 用のキーペアの作成と公開鍵の登録
~/devstack$ nova boot --image ubuntu --flavor m1.small --key mykey vm01       # vm01
を作成
~/devstack$ nova boot --image ubuntu --flavor m1.small --key mykey vm-mirror  #
vm-mirror を作成
~/devstack$ neutron floatingip-create ext-net                                 # フロー
ティング IP の作成
Created a new floatingip:
+---------------------+--------------------------------------+
| Field               | Value                                |
+---------------------+--------------------------------------+
| fixed_ip_address    |                                      |
| floating_ip_address | 172.16.1.4                           |
| floating_network_id | ad490f39-82b9-4164-8526-d47ae379dd49 |
| id                  | 101351bd-3044-4da2-bf36-72a193adc976 |
| port_id             |                                      |
| router_id           |                                      |
| status              | DOWN                                 |
| tenant_id           | 48e3c8d0688c4ccfac7126060ec550c0     |
+---------------------+--------------------------------------+
~/devstack$ neutron floatingip-create ext-net                                 # フロー
ティング IP の作成
Created a new floatingip:
+---------------------+--------------------------------------+
| Field               | Value                                |
+---------------------+--------------------------------------+
| fixed_ip_address    |                                      |
| floating_ip_address | 172.16.1.5                           |
| floating_network_id | ad490f39-82b9-4164-8526-d47ae379dd49 |
| id                  | f75ba7ed-3796-4f97-a0b9-e39d77baafca |
| port_id             |                                      |
| router_id           |                                      |
| status              | DOWN                                 |
| tenant_id           | 48e3c8d0688c4ccfac7126060ec550c0     |
+---------------------+--------------------------------------+
~/devstack$ neutron port-list                                                 # インス
```

```
タンスのポートを確認
+--------------------------------------+------+-------------------+---------------------------------------------------------------------------------+
| id                                   | name | mac_address       | fixed_ips                                                                       |
+--------------------------------------+------+-------------------+---------------------------------------------------------------------------------+
| 18173079-b4b1-44f8-85cc-a8bdd2a9ebe4 |      | fa:16:3e:db:c3:fd | {"subnet_id": "565fdfc0-d215-46c1-88da-fdd938c3aebd", "ip_address": "10.11.12.8"} |
| 470bbdfe-4be4-49c0-8a0b-71a75f664e9c |      | fa:16:3e:57:9e:07 | {"subnet_id": "565fdfc0-d215-46c1-88da-fdd938c3aebd", "ip_address": "10.11.12.2"} |
| c22dc6bc-d575-4100-a825-c64ef3a2fac4 |      | fa:16:3e:da:ac:59 | {"subnet_id": "565fdfc0-d215-46c1-88da-fdd938c3aebd", "ip_address": "10.11.12.1"} |
| d151ef5d-5954-4c35-ac88-b047f6bda2b1 |      | fa:16:3e:c5:8b:b9 | {"subnet_id": "565fdfc0-d215-46c1-88da-fdd938c3aebd", "ip_address": "10.11.12.7"} |
+--------------------------------------+------+-------------------+---------------------------------------------------------------------------------+
~/devstack$ neutron floatingip-associate 101351bd-3044-4da2-bf36-72a193adc976 d151ef5d-5954-4c35-ac88-b047f6bda2b1   # vm01 へフローティング IP を割り当てる
~/devstack$ neutron floatingip-associate f75ba7ed-3796-4f97-a0b9-e39d77baafca 18173079-b4b1-44f8-85cc-a8bdd2a9ebe4   # vm-mirror へフローティング IP を割り当てる
```

コンピュートノードの CLI を 2 つ開き、Floating IP で vm01 と vm-mirror に SSH ログインします。各インスタンスで eth0 への ICMP パケットを tcpdump で監視します。

```
[vm01]
```bash
~/devstack$ ssh 172.16.1.4 -i mykey
Welcome to Ubuntu 14.04.4 LTS (GNU/Linux 3.13.0-85-generic x86_64)

 * Documentation:  https://help.ubuntu.com/


  System information disabled due to load higher than 1.0




  Get cloud support with Ubuntu Advantage Cloud Guest:
    http://www.ubuntu.com/business/services/cloud




0 packages can be updated.
0 updates are security updates.




The programs included with the Ubuntu system are free software;
the exact distribution terms for each program are described in the
individual files in /usr/share/doc/*/copyright.




Ubuntu comes with ABSOLUTELY NO WARRANTY, to the extent permitted by
```
```

第 4 章　RESTCONF API を使ってフローを書き換えてみる

```
applicable law.


ubuntu@vm01:~$ sudo tcpdump -i eth0 icmp
tcpdump: verbose output suppressed, use -v or -vv for full protocol decode
listening on eth0, link-type EN10MB (Ethernet), capture size 65535 bytes


//emlist{

[vm-mirror]
'''bash
~/devstack$ ssh 172.16.1.5 -i mykey
Welcome to Ubuntu 14.04.4 LTS (GNU/Linux 3.13.0-85-generic x86_64)

 * Documentation:  https://help.ubuntu.com/

 System information disabled due to load higher than 1.0

  Get cloud support with Ubuntu Advantage Cloud Guest:
    http://www.ubuntu.com/business/services/cloud

0 packages can be updated.
0 updates are security updates.



The programs included with the Ubuntu system are free software;
the exact distribution terms for each program are described in the
individual files in /usr/share/doc/*/copyright.

Ubuntu comes with ABSOLUTELY NO WARRANTY, to the extent permitted by
applicable law.

ubuntu@vm-mirror:~$ sudo tcpdump -i eth0 icmp
tcpdump: verbose output suppressed, use -v or -vv for full protocol decode
listening on eth0, link-type EN10MB (Ethernet), capture size 65535 bytes
```

コントローラノードから vm01 の Floating IP に対して ping を実行します。

```
~/devstack$ ping 172.16.1.4
PING 172.16.1.4 (172.16.1.4) 56(84) bytes of data.
64 bytes from 172.16.1.4: icmp_seq=1 ttl=63 time=4.01 ms
64 bytes from 172.16.1.4: icmp_seq=2 ttl=63 time=5.66 ms
64 bytes from 172.16.1.4: icmp_seq=3 ttl=63 time=1.04 ms
64 bytes from 172.16.1.4: icmp_seq=4 ttl=63 time=1.00 ms
 :
```

vm01 の tcpdump で icmp リクエストが届いている事がわかると思います。

```
ubuntu@vm01:~$ sudo tcpdump -i eth0 icmp
tcpdump: verbose output suppressed, use -v or -vv for full protocol decode
listening on eth0, link-type EN10MB (Ethernet), capture size 65535 bytes
08:58:25.002574 IP 172.16.1.1 > host-10-11-12-3.openstacklocal: ICMP echo request,
id 9565, seq 1, length 64
08:58:25.004569 IP host-10-11-12-3.openstacklocal > 172.16.1.1: ICMP echo reply, id
9565, seq 1, length 64
08:58:26.018485 IP 172.16.1.1 > host-10-11-12-3.openstacklocal: ICMP echo request,
id 9565, seq 2, length 64
08:58:26.018673 IP host-10-11-12-3.openstacklocal > 172.16.1.1: ICMP echo reply, id
9565, seq 2, length 64
08:58:27.017110 IP 172.16.1.1 > host-10-11-12-3.openstacklocal: ICMP echo request,
id 9565, seq 3, length 64
08:58:27.017286 IP host-10-11-12-3.openstacklocal > 172.16.1.1: ICMP echo reply, id
9565, seq 3, length 64
08:58:28.018369 IP 172.16.1.1 > host-10-11-12-3.openstacklocal: ICMP echo request,
id 9565, seq 4, length 64
08:58:28.018548 IP host-10-11-12-3.openstacklocal > 172.16.1.1: ICMP echo reply, id
9565, seq 4, length 64
:
```

それでは、vm01 宛のパケットを vm-mirror へミラーリングためのフローを設定します。

```
~/devstack$ odl flow create openflow:
267140174830659 110 flow_mirror --tun-id 0x420 --dl-dst fa:16:3e:c5:8b:b9
--instructions output:2,output:3
True
```

curl でリクエストを投げると以下の様になります。

```
~/devstack$ curl -i -H
"Content-Type: application/xml" -H "Accept: application/json" -u "admin:admin" -X PUT
"http://192.168.0.11:8181/restconf/config/opendaylight-inventory:nodes/node/openflow:
267140174830659/flow-node-inventory:table/110/flow/flow_mirror" -d '<?xml
version="1.0" encoding="utf-8"?><flow
xmlns="urn:opendaylight:flow:inventory"><table_id>110</table_id><id>flow_mirror</id>
<match><tunnel><tunnel-id>0x420</tunnel-id></tunnel><ethernet-match><ethernet-desti
nation><mask>ff:ff:ff:ff:ff:ff</mask><address>fa:16:3e:c5:8b:b9</address></ethernet-
destination></ethernet-match></match><instructions><instruction><order>0</order><app
ly-actions><action><output-action><output-node-connector>2</output-node-connector></
output-action><order>0</order></action><action><output-action><output-node-connector>
3</output-node-connector></output-action><order>1</order></action></apply-actions></
instruction></instructions></flow>'
```

vm-mirror の tcpdump を見てみるとミラーリングされたパケットが届くようになったことが確認できます。

```
[vm-mirror]
@<tt>{bash
ubuntu@vm-mirror:~$ sudo tcpdump -i eth0 icmp
```

第 4 章　RESTCONF API を使ってフローを書き換えてみる

```
tcpdump: verbose output suppressed, use -v or -vv for full protocol decode
listening on eth0, link-type EN10MB (Ethernet), capture size 65535 bytes
09:07:04.530341 IP 172.16.1.1 > 10.11.12.3: ICMP echo request, id 9788, seq 1,
 length 64
09:07:05.533755 IP 172.16.1.1 > 10.11.12.3: ICMP echo request, id 9788, seq 2,
 length 64
09:07:06.531791 IP 172.16.1.1 > 10.11.12.3: ICMP echo request, id 9788, seq 3,
 length 64
    :
```

　RESTCONF API を使ってミラーリングのフローを設定する例を紹介しました。OpenDaylight が MD-SAL という抽象化レイヤーを採用していることによって、OpenDaylight を利用する様々なアプリケーションからのリクエストを NBI から受け取れる仕組みになっています。本章は RESTCONF API をご紹介しましたが、この他にも様々な NBI が用意されているので、ぜひ使ってみてください。

第5章 OpenStack TackerによるNFVオーケストレーション

　OpenDaylight のユースケースに NFV（Network Functions Virtualization）があります。本章は NFV の視点から、Tacker（https://wiki.openstack.org/wiki/Tacker）について解説します。Tacker とは、OpenStack のプロジェクトの1つで、ETSI MANO アーキテクチャフレームワークに基づいた VNFM/NFVO の役割を果たすコンポーネントです。といっても、VNFM/NFVO って何？という方も多いと思いますので、ETSI（欧州電気通信標準化機構）が提唱する NFV について簡単におさらいしておきます。

5.1　NFVアーキテクチャのおさらい

　NFV とは、ファイアーウォールなどのこれまで専用のハードウェアアプライアンスで実現されていたネットワーク機能を、ソフトウェア（仮想マシン）として実現しようというものです。そのソフトウェアで動作するネットワーク機能を VNF（Virtualized Network Function）と呼びます。そして、VNF の実行基盤に必要なストレージやネットワークなどのハードウェアやハイパーバイザーなどのソフトウェアを NFVI（Network Functions Virtualization Infrastructure）と呼びます。

　そうなると、VNF や NFVI をまとめて管理するコンポーネントが必要になってきます。それが NFV-MANO（NFV Management and Orchestration）になります。NFV-MANO は以下に示す3つの機能ブロックを持ちます。

- VIM（Virtualized Infrastructure Manager）
    - NFVI のコンピューティング、ストレージおよびネットワークリソースの制御、管理

を行います
- NFVO（NFV Orchestrator）
  - NFVI のオーケストレーションやネットワークサービスのライフサイクルの管理を行います
- VNFM（VNF Manager）
  - VNF の設定やテンプレートによる展開、ソフトウェアアップグレードの管理、イベント/障害検出、オートヒーリングなど、VNF インスタンスのライフサイクルの管理を行います

OpenStack で NFV を実現するために Tacker が担っているのは、これらの機能ブロックの中の NFVO と VNFM です。そして、将来的には Tacker は Neutron の networking-sfc を経由して OpenDaylight（以降、ODL と省略）を利用する仕組みになります。この Tacker と ODL 連携の仕組みは、2016 年 4 月現在、実装が進められている段階ですが、すでに各方面から注目されており、ODL を用いた NFV として広く使われる可能性があります。

しかしながら、Tacker については、OpenDaylight 以上に、構築についての情報が少なく、手動による構築は難易度が高いです。前回まで、OpenStack 環境の構築に DevStack を利用しました。そこで、今回も DevStack を利用して Tacker を構築した後、VNF を作成して使ってみるところまで解説していきます。

5.2 環境・バージョンについて

今回の構築はシングルノードの構成とします。筆者の環境では vSphere ESXi 上に Nested KVM を有効にした Ubuntu の仮想マシンを 1 台用意しています。この仮想マシンは、DevStack の性質上、ルート権限（sudo）での実行が可能である必要があります。さらに、ソースコードのダウンロードを行うためインターネットアクセスができる環境が必要です。

OS ディストリビューション	カーネルバージョン
Ubuntu 14.04.2 LTS	Linux version 3.16.0-30-generic

OpenStack 環境の各コンポーネントは Mitaka を使用します。

コンポーネント	バージョン
Nova	
Glance	
Neutron	
Horizon	
Tacker	Mitaka

5.3 構築してみよう

それでは構築を始めていきます。まずは、構築を簡単にするためにファイアーウォールを無効化しておきます。

```
~$ sudo ufw disable
```

次に DevStack をダウンロードします。その際、git コマンドが必要なので必要に応じてインストールしてください。ブランチ名は stable/mitaka を指定します。

```
~$ sudo apt-get install git
~$ git clone --branch stable/mitaka https://github.com/openstack-dev/devstack.git
~$ cd devstack/
```

今回使用する local.conf をダウンロードします。

```
~/devstack$ wget https://raw.githubusercontent.com/YujiAzama/openstack-tacker/master/local.conf
```

後は、stack.sh スクリプトを実行するだけです。以下の様なメッセージが出力されれば終了です。

```
~/devstack$ ./stack.sh
    :
This is your host IP address: 192.168.0.10
This is your host IPv6 address: ::1
Horizon is now available at http://192.168.0.10/dashboard
Keystone is serving at http://192.168.0.10:5000/
The default users are: admin and demo
The password: password
~/devstack$
```

以上で構築作業は終了です。

デフォルトで net_mgmt、external、net0、net1 のネットワークとそれぞれのサブネットが作成されていることを確認します。

第 5 章　OpenStack Tacker による NFV オーケストレーション

```
~/devstack$ source openrc admin admin
WARNING: setting legacy OS_TENANT_NAME to support cli tools.
~/devstack$ neutron net-list
+--------------------------------------+----------+------------------------------------
| id                                   | name     | subnets
|
+--------------------------------------+----------+------------------------------------
| 33afc037-bcb9-4fa0-bd90-ca54f1bade8f | external |
f631a6b5-7f58-4a17-ae70-4e715cedbbdf 2001:db8::/64          |
|                                                                                     
556f45ec-2f62-4909-94ea-2428552fc891 10.12.161.0/24         |
| d1aa4aa3-9fb0-47b3-afa2-4af17d68e7ed | net1     |
b141ef9d-610c-4e0f-b7a9-6a60cd2ce305 10.10.1.0/24           |
| 45b92be3-0a30-4df5-a2eb-38aba23b4da0 | net0     |
b15d2770-ee41-49ca-9ce9-1fdd8e6fa2c8 10.10.0.0/24           |
| 6208b02e-130a-4622-a228-d73c27ddd587 | private  |
92409cb3-f8a2-4528-a2ee-2cb324b3a8e0 fd9f:c646:8f20::/64    |
|                                                                                     
a0f4af9b-8493-46fe-b776-79f4828fe870 15.0.0.0/24            |
| 71c9ae3b-331d-4dbf-b090-cf53a9b177ce | net_mgmt |
c5219e79-bed8-4f6e-8c68-49fbf8b2c483 192.168.120.0/24       |
+--------------------------------------+----------+------------------------------------
~/devstack$ neutron subnet-list
+--------------------------------------+--------------------+---------------------+---------------------+
| id                                   | name               | cidr                |
allocation_pools                                                                                         |
+--------------------------------------+--------------------+---------------------+---------------------+
| 556f45ec-2f62-4909-94ea-2428552fc891 | public-subnet      | 10.12.161.0/24      |
{"start": "10.12.161.150", "end": "10.12.161.201"}                                                       |
| f631a6b5-7f58-4a17-ae70-4e715cedbbdf | ipv6-public-subnet | 2001:db8::/64       |
{"start": "2001:db8::3", "end": "2001:db8::ffff:ffff:ffff:ffff"}                                         |
|                                      |                    |                     |
{"start": "2001:db8::1", "end": "2001:db8::1"}                                                           |
| 92409cb3-f8a2-4528-a2ee-2cb324b3a8e0 | ipv6-private-subnet| fd9f:c646:8f20::/64 |
{"start": "fd9f:c646:8f20::2", "end": "fd9f:c646:8f20:0:ffff:ffff:ffff:ffff"}                            |
| a0f4af9b-8493-46fe-b776-79f4828fe870 | private-subnet     | 15.0.0.0/24         |
{"start": "15.0.0.2", "end": "15.0.0.254"}                                                               |
| b141ef9d-610c-4e0f-b7a9-6a60cd2ce305 | subnet1            | 10.10.1.0/24        |
{"start": "10.10.1.2", "end": "10.10.1.254"}                                                             |
| b15d2770-ee41-49ca-9ce9-1fdd8e6fa2c8 | subnet0            | 10.10.0.0/24        |
{"start": "10.10.0.2", "end": "10.10.0.254"}                                                             |
| c5219e79-bed8-4f6e-8c68-49fbf8b2c483 | subnet_mgmt        | 192.168.120.0/24    |
{"start": "192.168.120.2", "end": "192.168.120.254"}                                                     |
+--------------------------------------+--------------------+---------------------+---------------------+
```

構築した OpenStack が、VIM として登録されていることも確認します。

```
~/devstack$ tacker vim-list
+--------------------------------------+-----------------------------------+------+--
| id                                   | tenant_id                         | name |
type         | description | auth_url  | placement_attr                    |
```

```
  auth_cred                                                                                                     |
+----------------------------------+----------------------------------+------+--
| 1cdd90eb-c147-438a-8764-52cef1f4a0e8 | 3eda2401d7eb40c0bcdf655d1218ec62 | VIM0 |
openstack |                       | http://localhost:5000/v3 | {u'regions': [u'RegionOne']} |
{u'username': u'nfv_user', u'project_name': u'nfv',       |
|                       |                       |                       |       |
|                       |                       |                       | u'user_id':
u'', u'user_domain_id': u'default',           |
|                       |                       |                       |       |
|                       |                       |                       |       |
u'project_id': u'', u'auth_url':                     |
|                       |                       |                       |       |
|                       |                       |                       |       |
u'http://localhost:5000/v3', u'password': '***',     |
|                       |                       |                       |       |
|                       |                       |                       |       |
u'project_domain_id': u'default'}                    |
+----------------------------------+----------------------------------+------+--
```

5.4 TackerでVNFをデプロイしてみよう

　VNFを作成するには、先にVNFD（VNF Descriptor）を作成する必要があります。VNFDとはTOSCAで定義されたVNFのトポロジに対して、どのようにデプロイやオペレーションするかを定義したものです。TOSCA（Topology and Orchestration Specification for Cloud Applications）とは、クラウド上に展開されるシステム構成（トポロジ）を定義したものです。TOSCAを使用することによって、異なるベンダ間のクラウドアプリケーションおよびサービスをスムーズに導入でき、多種多様なクラウドをまたがったアプリケーションの可搬性を確保し、システムの移行や連携を容易にできるメリットが得られます。

　今回使用するTOSCAテンプレートの中では、VNFインスタンスのイメージの指定の他、net0とnet1、net_mgmtという3つのネットワークをVNFインスタンスに接続するように定義しています。

VNFイメージの登録

　今回使用するVNFは、ファイアーウォールに見立てた簡易的なもので、Ubuntuのイメージをベースに筆者が独自に作成しました。ファイアーウォール機能にはiptablesを使っています。まず、そのVNFイメージのダウンロードとGlanceへの登録を行います。

```
~/devstack$ wget -O myvnf.qcow2
https://www.googledrive.com/host/0Bx3-Oe8aeU3FOUwxVlBLYW5KNEU
  :
myvnf.qcow2             [<=>              ] 373.19M  10.9MB/s    in 34s

2016-05-09 13:46:13 (10.9 MB/s) - 'myvnf.qcow2' へ保存終了 [391315456]

~/devstack$ glance image-create --file myvnf.qcow2 --disk-format qcow2
--container-format bare --name='myvnf' --visibility public
+------------------+--------------------------------------+
| Property         | Value                                |
+------------------+--------------------------------------+
| checksum         | bbf3134efa2f67d31b48e2f681a37198     |
| container_format | bare                                 |
| created_at       | 2016-05-16T01:35:19Z                 |
| disk_format      | qcow2                                |
| id               | 876a2de8-58b5-41e3-8f53-e21de6f39098 |
| min_disk         | 0                                    |
| min_ram          | 0                                    |
| name             | myvnf                                |
| owner            | 3eda2401d7eb40c0bcdf655d1218ec62     |
| protected        | False                                |
| size             | 391315456                            |
| status           | active                               |
| tags             | []                                   |
| updated_at       | 2016-05-16T01:35:22Z                 |
| virtual_size     | None                                 |
| visibility       | public                               |
+------------------+--------------------------------------+
```

　このVNFイメージを起動するために、RAMが3GB、ディスクが20GB程度のフレーバーを用意します。myvnf_flavorという名前のフレーバーを作成します。TOSCAテンプレート内でフレーバーを指定しているので同じ名前で作成して下さい。

```
~/devstack$ nova flavor-create --is-public true myvnf_flavor auto 3072 20 1
+--------------------------------------+--------------+-----------+------+-----------+
| ID                                   | Name         | Memory_MB | Disk | Ephemeral |
| Swap | VCPUs | RXTX_Factor | Is_Public |
+--------------------------------------+--------------+-----------+------+-----------+
| d4c11af1-0d00-4a87-bcb0-ebf3f52b32e9 | myvnf_flavor | 3072      | 20   | 0         |
|      | 1     | 1.0         | True       |
+--------------------------------------+--------------+-----------+------+-----------+
```

VNFDの作成

　それではVNFDを作成しますが、先にTOSCAテンプレートをダウンロードします。

5.4 Tacker で VNF をデプロイしてみよう

```
~/devstack$ wget
https://raw.githubusercontent.com/YujiAzama/openstack-tacker/master/myvnf.yaml
~/devstack$ tacker vnfd-create --vnfd-file myvnf.yaml
Created a new vnfd:
+---------------+------------------------------------------------------------
| Field         | Value                                                                    |
+---------------+------------------------------------------------------------
| description   | A simple firewall based on Ubuntu.                                       |
| id            | 1b909041-76c8-4344-b10f-bff3d484973e                                     |
| infra_driver  | heat                                                                     |
| mgmt_driver   | noop                                                                     |
| name          | MyVNF                                                                    |
| service_types | {"service_type": "vnfd", "id": "7ed83b08-c658-4ffa-9b0b-4861290b1f7f"}   |
| tenant_id     | 3eda2401d7eb40c0bcdf655d1218ec62                                         |
+---------------+------------------------------------------------------------
```

VNF の作成

先ほど作成した VNFD の ID を指定して VNF を作成します。完了するまで、1 分程度かかる場合があります。

```
~/devstack$ tacker vnf-create --name firewall --vnfd-id
1b909041-76c8-4344-b10f-bff3d484973e
Created a new vnf:
+----------------+------------------------------------------------+
| Field          | Value                                          |
+----------------+------------------------------------------------+
| description    | A simple firewall based on Ubuntu.             |
| id             | 742e0260-115e-4832-8613-269706fd75e1           |
| instance_id    | 966bc8f8-4a02-41b9-b241-99f1c7d7ce75           |
| mgmt_url       |                                                |
| name           | firewall                                       |
| placement_attr | {"vim_name": "VIM0"}                           |
| status         | PENDING_CREATE                                 |
| tenant_id      | 3eda2401d7eb40c0bcdf655d1218ec62               |
| vim_id         | 1cdd90eb-c147-438a-8764-52cef1f4a0e8           |
| vnfd_id        | 1b909041-76c8-4344-b10f-bff3d484973e           |
+----------------+------------------------------------------------+
```

以下の様に、作成した VNF のステータスが「PENDING_CREATE」から「ACTIVE」になれば作成完了です。

```
~/devstack$ tacker vnf-list
+--------------------------------------+----------+----------------------------------
| id                                   | name     | description
| mgmt_url              | status | vim_id                                           |
placement_attr         |
+--------------------------------------+----------+----------------------------------
| 742e0260-115e-4832-8613-269706fd75e1 | firewall | A simple firewall based on
Ubuntu. | {"vdu1": "192.168.120.3"} | ACTIVE | 1cdd90eb-c147-438a-8764-52cef1f4a0e8
| {u'vim_name': u'VIM0'} |
+--------------------------------------+----------+----------------------------------
```

5.5　VNFインスタンスはserviceテナントに作られる

NovaコマンドからVNFインスタンスの状態を確認してみます。

```
~/devstack$ nova list
+----+------+--------+------------+-------------+----------+
| ID | Name | Status | Task State | Power State | Networks |
+----+------+--------+------------+-------------+----------+
+----+------+--------+------------+-------------+----------+
```

先ほど作成したVNFインスタンスが見当たりません。実は、tackerによって展開されるVNFインスタンスはserviceテナントに作成されます。--all-tenantオプションを付けると全テナントのインスタンスが取得できます。

```
~/devstack$ nova list --all-tenant
+--------------------------------------+-----------------------------------------
| ID                                   | Name
| Tenant ID                            | Status | Task State | Power State | Networks
|
+--------------------------------------+-----------------------------------------
| 0d2f79c0-149d-4f18-81fa-1c8810a274e1 |
ta-0260-115e-4832-8613-269706fd75e1-vdu1-5xyg3hv6icwe |
cb67d4395179494fa730979ccf73dcf3 | ACTIVE | -          | Running     |
net_mgmt=192.168.120.3; net1=10.10.1.3; net0=10.10.0.3 |
+--------------------------------------+-----------------------------------------
```

5.6　VNFの構成

今回作成したVNFインスタンスはnet0とnet1およびnet_mgmtのネットワークに接続されています。

5.7 VNFを使ってみる

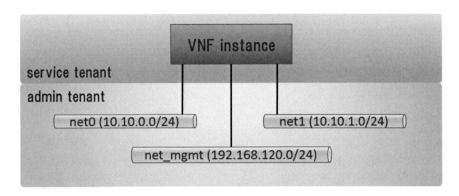

図 5.1　ネットワーク図

net0 は VNF インスタンスへのインバウンドパケットのためのネットワーク、それに対して net1 は VNF インスタンスからのアウトバウンドパケットのためのネットワークです。net_mgmt というネットワークは VNF インスタンスのための管理用ネットワークです。管理用ネットワークは、VNF インスタンスの設定やヘルスチェックなどを行うためのネットワークなので、全ての VNF インスタンスが接続されることになります。VNF の構成は TOSCA テンプレートで記述すると説明しましたが、今回使用した TOSCA テンプレートの中で、VNF インスタンスのイメージや接続するネットワーク、ヘルスチェックなど、VNF の構成を定義しています。

5.7　VNFを使ってみる

では、実際に VNF インスタンスを使ってみます。先ほど作成した VNF は iptables を使った簡易的なファイアーウォール機能として動作します。

VNF インスタンスにはネットワーク net0 と net1 が接続されていました。そこで、それぞれのネットワークにクライアントとなるインスタンス net0-vm（cirros）と net1-vm（cirros）を作成し疎通確認をしてみます。インスタンスは admin テナントで作成します。疎通確認は次の 2 点を確認します。

1. 初期状態の VNF の設定では net0 から net1 宛の ping が到達すること
2. VNF（iptables）を設定することによって、net0 から net1 宛の ping がブロックされるようになること

第 5 章　OpenStack Tacker による NFV オーケストレーション

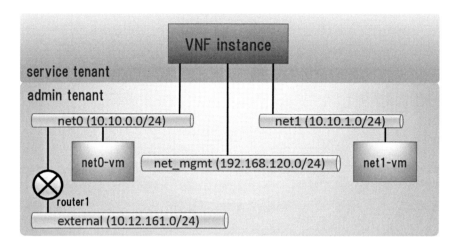

図 5.2　ネットワーク図

経路情報の追加

　net0 と net1 のインスタンス間を VNF 経由で相互に通信できるように経路情報を登録する必要があります。まず、VNF の net0 側と net1 側の IP を調べます。

```
~/devstack$ nova list --all-tenant
+--------------------------------------+------------------------------------------
| ID                                   | Name
| Tenant ID                            | Status | Task State | Power State | Networks
|
+--------------------------------------+------------------------------------------
| 0d2f79c0-149d-4f18-81fa-1c8810a274e1 |
ta-0260-115e-4832-8613-269706fd75e1-vdu1-5xyg3hv6icwe |
cb67d4395179494fa730979ccf73dcf3 | ACTIVE |  -         | Running     |
net_mgmt=192.168.120.3; net1=10.10.1.3; net0=10.10.0.3 |
+--------------------------------------+------------------------------------------
```

　経路情報を登録します。

```
~/devstack$ neutron subnet-update subnet0 --host-route
destination=10.10.1.0/24,nexthop=10.10.0.3
Updated subnet: subnet0
~/devstack$ neutron subnet-update subnet1 --host-route
destination=10.10.0.0/24,nexthop=10.10.1.3
Updated subnet: subnet1
```

インスタンスの作成

　それでは、net0-vm と net1-vm を作成します。ping による疎通確認さえできれば良いので

5.7 VNF を使ってみる

cirros イメージを使用します。external ネットワークからインスタンスに SSH 接続出きるようにフローティング IP も割り当てます。

```
~/devstack$ neutron net-list    # ネットワークの ID を調べる
+--------------------------------------+---------+----------------------------------+
| id                                   | name    | subnets                          |
+--------------------------------------+---------+----------------------------------+
| 45b92be3-0a30-4df5-a2eb-38aba23b4da0 | net0    |
b15d2770-ee41-49ca-9ce9-1fdd8e6fa2c8 10.10.0.0/24       | # net0
| 6208b02e-130a-4622-a228-d73c27ddd587 | private |
92409cb3-f8a2-4528-a2ee-2cb324b3a8e0 fd9f:c646:8f20::/64 |
|                                      |         |
a0f4af9b-8493-46fe-b776-79f4828fe870 15.0.0.0/24        |
| 71c9ae3b-331d-4dbf-b090-cf53a9b177ce | net_mgmt |
c5219e79-bed8-4f6e-8c68-49fbf8b2c483 192.168.120.0/24   |
| 33afc037-bcb9-4fa0-bd90-ca54f1bade8f | external |
556f45ec-2f62-4909-94ea-2428552fc891 10.12.161.0/24     |
|                                      |         |
f631a6b5-7f58-4a17-ae70-4e715cedbbdf 2001:db8::/64      |
| d1aa4aa3-9fb0-47b3-afa2-4af17d68e7ed | net1    |
b141ef9d-610c-4e0f-b7a9-6a60cd2ce305 10.10.1.0/24       | # net1
+--------------------------------------+---------+----------------------------------+
~/devstack$ nova boot --image cirros-0.3.4-x86_64-uec --flavor m1.tiny --nic
net-id=45b92be3-0a30-4df5-a2eb-38aba23b4da0 net0-vm   # net0 のネットワークを指定する
+--------------------------------------+------------------------------------------+
| Property                             | Value                                    |
+--------------------------------------+------------------------------------------+
| OS-DCF:diskConfig                    | MANUAL                                   |
| OS-EXT-AZ:availability_zone          |                                          |
| OS-EXT-SRV-ATTR:host                 | -                                        |
| OS-EXT-SRV-ATTR:hostname             | net0-vm                                  |
| OS-EXT-SRV-ATTR:hypervisor_hostname  | -                                        |
| OS-EXT-SRV-ATTR:instance_name        | instance-00000002                        |
| OS-EXT-SRV-ATTR:kernel_id            | d208b1b8-815f-4916-a0f8-b6a0f9904dfe     |
| OS-EXT-SRV-ATTR:launch_index         | 0                                        |
| OS-EXT-SRV-ATTR:ramdisk_id           | 257c763b-5457-47c7-98a8-ff6b9455a9be     |
| OS-EXT-SRV-ATTR:reservation_id       | r-nutjal63                               |
| OS-EXT-SRV-ATTR:root_device_name     | -                                        |
| OS-EXT-SRV-ATTR:user_data            | -                                        |
```

第 5 章　OpenStack Tacker による NFV オーケストレーション

```
|                                      |                                       |
| OS-EXT-STS:power_state                | 0                                     |
| OS-EXT-STS:task_state                 | scheduling                            |
| OS-EXT-STS:vm_state                   | building                              |
| OS-SRV-USG:launched_at                | -                                     |
| OS-SRV-USG:terminated_at              | -                                     |
| accessIPv4                            |                                       |
| accessIPv6                            |                                       |
| adminPass                             | cJm8nfCq26Uz                          |
| config_drive                          |                                       |
| created                               | 2016-05-16T01:45:29Z                  |
| description                           | -                                     |
| flavor                                | m1.tiny (1)                           |
| hostId                                |                                       |
| host_status                           |                                       |
| id                                    | fbc05daa-a1b9-4015-91e8-94ac02109d09  |
| image                                 | cirros-0.3.4-x86_64-uec               |
(25db4eab-f79f-4773-8405-48e1da9dd543) |
| key_name                              | -                                     |
| locked                                | False                                 |
| metadata                              | {}                                    |
| name                                  | net0-vm                               |
| os-extended-volumes:volumes_attached  | []                                    |
| progress                              | 0                                     |
| security_groups                       | default                               |
| status                                | BUILD                                 |
| tenant_id                             | 3eda2401d7eb40c0bcdf655d1218ec62      |
| updated                               | 2016-05-16T01:45:30Z                  |
```

5.7 VNF を使ってみる

```
| user_id                              | 15c5a599501e43abac606f643c1a7a6f          |
+--------------------------------------+-------------------------------------------+
~/devstack$ nova boot --image cirros-0.3.4-x86_64-uec --flavor m1.tiny --nic
net-id=d1aa4aa3-9fb0-47b3-afa2-4af17d68e7ed net1-vm   # net1 のネットワークを指定する
+--------------------------------------+-------------------------------------------+
| Property                             | Value                                     |
+--------------------------------------+-------------------------------------------+
| OS-DCF:diskConfig                    | MANUAL                                    |
| OS-EXT-AZ:availability_zone          |                                           |
| OS-EXT-SRV-ATTR:host                 | -                                         |
| OS-EXT-SRV-ATTR:hostname             | net1-vm                                   |
| OS-EXT-SRV-ATTR:hypervisor_hostname  | -                                         |
| OS-EXT-SRV-ATTR:instance_name        | instance-00000003                         |
| OS-EXT-SRV-ATTR:kernel_id            | d208b1b8-815f-4916-a0f8-b6a0f9904dfe      |
| OS-EXT-SRV-ATTR:launch_index         | 0                                         |
| OS-EXT-SRV-ATTR:ramdisk_id           | 257c763b-5457-47c7-98a8-ff6b9455a9be      |
| OS-EXT-SRV-ATTR:reservation_id       | r-1ddzn7wd                                |
| OS-EXT-SRV-ATTR:root_device_name     | -                                         |
| OS-EXT-SRV-ATTR:user_data            | -                                         |
| OS-EXT-STS:power_state               | 0                                         |
| OS-EXT-STS:task_state                | scheduling                                |
| OS-EXT-STS:vm_state                  | building                                  |
| OS-SRV-USG:launched_at               | -                                         |
| OS-SRV-USG:terminated_at             | -                                         |
| accessIPv4                           |                                           |
| accessIPv6                           |                                           |
| adminPass                            | 7NTvzYv6f7eK                              |
| config_drive                         |                                           |
| created                              | 2016-05-16T01:46:28Z                      |
```

第 5 章　OpenStack Tacker による NFV オーケストレーション

```
| description                        | -                                        |
| flavor                             | m1.tiny (1)                              |
| hostId                             |                                          |
| host_status                        |                                          |
| id                                 | bce5e7fd-d1b4-4cbc-8049-54a9e798fc9e     |
| image                              | cirros-0.3.4-x86_64-uec                  |
(25db4eab-f79f-4773-8405-48e1da9dd543) |                                       |
| key_name                           | -                                        |
| locked                             | False                                    |
| metadata                           | {}                                       |
| name                               | net1-vm                                  |
| os-extended-volumes:volumes_attached | []                                     |
| progress                           | 0                                        |
| security_groups                    | default                                  |
| status                             | BUILD                                    |
| tenant_id                          | 3eda2401d7eb40c0bcdf655d1218ec62         |
| updated                            | 2016-05-16T01:46:28Z                     |
| user_id                            | 15c5a599501e43abac606f643c1a7a6f         |
+------------------------------------+------------------------------------------+
~/devstack$ nova list   # 作成したインスタンスの IP アドレスを確認する
+--------------------------------------+---------+--------+------------+-------------+-----------------+
| ID                                   | Name    | Status | Task State | Power State | Networks        |
+--------------------------------------+---------+--------+------------+-------------+-----------------+
| fbc05daa-a1b9-4015-91e8-94ac02109d09 | net0-vm | ACTIVE | -          | Running     | net0=10.10.0.4  |
| bce5e7fd-d1b4-4cbc-8049-54a9e798fc9e | net1-vm | ACTIVE | -          | Running     | net1=10.10.1.4  |
+--------------------------------------+---------+--------+------------+-------------+-----------------+
~/devstack$ neutron router-interface-add router1 subnet0
Added interface 5425d90a-2d26-46a8-86a0-490744e7d877 to router router1.
~/devstack$ neutron floatingip-create external
Created a new floatingip:
+---------------------+--------------------------------------+
| Field               | Value                                |
+---------------------+--------------------------------------+
| description         |                                      |
```

5.7 VNF を使ってみる

```
| dns_domain          |                                      |
| dns_name            |                                      |
| fixed_ip_address    |                                      |
| floating_ip_address | 10.12.161.151                        |
| floating_network_id | 33afc037-bcb9-4fa0-bd90-ca54f1bade8f |
| id                  | c4a58b5f-e372-441d-889c-3eb18a89e9c8 |
| port_id             |                                      |
| router_id           |                                      |
| status              | DOWN                                 |
| tenant_id           | 3eda2401d7eb40c0bcdf655d1218ec62     |
+---------------------+--------------------------------------+
~/devstack$ neutron port-list
+--------------------------------------+------------------------------------------
| id                                   | name
| mac_address       | fixed_ips
|
+--------------------------------------+------------------------------------------
| 105338dc-6c30-49a8-a9b5-fab17f04e211 |
| fa:16:3e:f1:09:b2 | {"subnet_id": "b141ef9d-610c-4e0f-b7a9-6a60cd2ce305",
"ip_address": "10.10.1.2"}            |
| 1aadee40-9606-44ec-b1d7-dd917f58257b |
| fa:16:3e:fe:d0:05 | {"subnet_id": "92409cb3-f8a2-4528-a2ee-2cb324b3a8e0",
"ip_address": "fd9f:c646:8f20::1"}    |
| 268ccf93-3ec5-46e9-bc4f-a631614afd1d |
| fa:16:3e:22:0d:2c | {"subnet_id": "a0f4af9b-8493-46fe-b776-79f4828fe870",
"ip_address": "15.0.0.1"}             |
| 45334855-bb9c-41b5-b6a1-62906a4a9be6 |
| fa:16:3e:69:ca:e9 | {"subnet_id": "b15d2770-ee41-49ca-9ce9-1fdd8e6fa2c8",
"ip_address": "10.10.0.2"}            |
| 5425d90a-2d26-46a8-86a0-490744e7d877 |
| fa:16:3e:c7:4a:b9 | {"subnet_id": "b15d2770-ee41-49ca-9ce9-1fdd8e6fa2c8",
"ip_address": "10.10.0.1"}            |
| 5762b752-83a3-499d-badd-9cddb3ede6d5 |
| fa:16:3e:fd:3c:07 | {"subnet_id": "b15d2770-ee41-49ca-9ce9-1fdd8e6fa2c8",
"ip_address": "10.10.0.4"}            |               # net0-vm
| 58a1b424-474b-4f09-98f4-4867b590d0a5 |
tacker.vm.infra_drivers.heat.heat_DeviceHeat-742e0260-115e-4832-8613-269706fd75e1-
vdu1-net_mgmt-port-   | fa:16:3e:d3:9a:e6 | {"subnet_id":
"c5219e79-bed8-4f6e-8c68-49fbf8b2c483", "ip_address": "192.168.120.3"}
|
|                                      | inspycqjjnjn
|                   |
|
| 5f4168c1-cd4e-4cf5-9790-98946e80388c |
| fa:16:3e:2f:28:aa | {"subnet_id": "a0f4af9b-8493-46fe-b776-79f4828fe870",
"ip_address": "15.0.0.2"}             |
|                   |
|                   | {"subnet_id": "92409cb3-f8a2-4528-a2ee-2cb324b3a8e0",
"ip_address":                         |
|                   |
|                   | "fd9f:c646:8f20:0:f816:3eff:fe2f:28aa"}
|
| 726acc29-7f25-4054-a7a5-c747008e7db6 |
```

第 5 章 OpenStack Tacker による NFV オーケストレーション

```
| fa:16:3e:fa:1b:65 | {"subnet_id": "b15d2770-ee41-49ca-9ce9-1fdd8e6fa2c8",
"ip_address": "10.10.0.3"}                |
| 74236635-fa67-47f0-9673-297f84549e6c |
| fa:16:3e:23:3f:4f | {"subnet_id": "556f45ec-2f62-4909-94ea-2428552fc891",
"ip_address": "10.12.161.150"}            |
|                   |                                                       
|                   | {"subnet_id": "f631a6b5-7f58-4a17-ae70-4e715cedbbdf",
"ip_address": "2001:db8::3"}              |
| 8cd5a363-687b-4ecd-ab3b-743ba7a18a50 |
| fa:16:3e:25:c7:e3 | {"subnet_id": "556f45ec-2f62-4909-94ea-2428552fc891",
"ip_address": "10.12.161.151"}            |
|                   |                                                       
|                   | {"subnet_id": "f631a6b5-7f58-4a17-ae70-4e715cedbbdf",
"ip_address": "2001:db8::4"}              |
| a3db77a1-bc28-41db-80cc-23092ee2d62b |
| fa:16:3e:b5:96:ed | {"subnet_id": "b141ef9d-610c-4e0f-b7a9-6a60cd2ce305",
"ip_address": "10.10.1.4"}                |   # net1-vm
| ab355d02-9d58-471b-96f7-bf0d6539ff0f |
| fa:16:3e:6f:7d:f5 | {"subnet_id": "b141ef9d-610c-4e0f-b7a9-6a60cd2ce305",
"ip_address": "10.10.1.3"}                |
| e5f4b673-6fe8-4dc6-b44f-1071e0fba0ea |
| fa:16:3e:7d:ce:75 | {"subnet_id": "c5219e79-bed8-4f6e-8c68-49fbf8b2c483",
"ip_address": "192.168.120.2"}            |
+-------------------+----------------------------------------------------
~/devstack$ neutron floatingip-associate c4a58b5f-e372-441d-889c-3eb18a89e9c8
5762b752-83a3-499d-badd-9cddb3ede6d5   # net0-vm のポートにフローティング IP を割り当てる
Associated floating IP c4a58b5f-e372-441d-889c-3eb18a89e9c8
```

今回使用している VNF インスタンスは、初期状態で全てのパケットの転送を許可するようにしています。そのため、今の状態では net0 から net1 への通信はできるようになっています。

別のターミナルを開き、net0-vm に SSH ログインして net1-vm に ping 疎通ができることを確認してみます。フローティング IP から接続します。ユーザー名に「cirros」で、パスワードに「cubswin:)」を入力することでログインできます。

```
~/devstack$ ssh cirros@10.12.161.151
The authenticity of host '10.12.161.151 (10.12.161.151)' can't be established.
RSA key fingerprint is 32:6d:3a:ba:93:00:2d:95:e3:76:51:8c:c5:d7:a2:43.
Are you sure you want to continue connecting (yes/no)? yes
Warning: Permanently added '10.12.161.151' (RSA) to the list of known hosts.
cirros@10.12.161.151's password:     # 「cubswin:)」と入力する
$ ping 10.10.1.4 -c 5
PING 10.10.1.4 (10.10.1.4): 56 data bytes
64 bytes from 10.10.1.4: seq=0 ttl=63 time=7.145 ms
64 bytes from 10.10.1.4: seq=1 ttl=63 time=0.695 ms
64 bytes from 10.10.1.4: seq=2 ttl=63 time=0.450 ms
64 bytes from 10.10.1.4: seq=3 ttl=63 time=0.663 ms
64 bytes from 10.10.1.4: seq=4 ttl=63 time=0.483 ms

--- 10.10.1.4 ping statistics ---
```

```
5 packets transmitted, 5 packets received, 0% packet loss
round-trip min/avg/max = 0.450/1.887/7.145 ms
```

net0-vm から net1-vm への疎通が確認できました。それでは、VNF の設定を変更して net0 から net1 へのパケットをブロックします。

VNF の設定はマネジメントネットワーク net_mgmt から SSH 接続して行います。まず、VNF のマネジメントネットワークの IP アドレスを確認します。

```
~/devstack$ tacker vnf-list    # VNF のマネジメントネットワーク IP アドレス (mgmt_url) を確認する
+--------------------------------------+----------+----------------------------------+
| id                                   | name     | description                      |
| mgmt_url                             | status   | vim_id                           |
| placement_attr                       |          |                                  |
+--------------------------------------+----------+----------------------------------+
| 742e0260-115e-4832-8613-269706fd75e1 | firewall | A simple firewall based on
Ubuntu. | {"vdu1": "192.168.120.3"} | ACTIVE | 1cdd90eb-c147-438a-8764-52cef1f4a0e8
| {u'vim_name': u'VIM0'} |
+--------------------------------------+----------+----------------------------------+
```

ユーザ名に「user」、パスワードに「password」を入力して VNF に SSH ログインします。

```
~/devstack$ ssh user@192.168.120.3
The authenticity of host '192.168.120.3 (192.168.120.3)' can't be established.
ECDSA key fingerprint is 31:3f:a9:9e:91:78:03:83:88:30:0c:93:5c:eb:19:11.
Are you sure you want to continue connecting (yes/no)? yes
Warning: Permanently added '192.168.120.3' (ECDSA) to the list of known hosts.
user@192.168.120.3's password:     # 「password」と入力する
Welcome to Ubuntu 14.04.4 LTS (GNU/Linux 3.13.0-85-generic x86_64)

 * Documentation:  https://help.ubuntu.com/

  System information as of Mon May 16 01:39:00 UTC 2016

  System load: 1.19              Memory usage: 2%    Processes:         102
  Usage of /:  67.1% of 1.42GB   Swap usage:   0%    Users logged in:   0

  Graph this data and manage this system at:
    https://landscape.canonical.com/

  Get cloud support with Ubuntu Advantage Cloud Guest:
    http://www.ubuntu.com/business/services/cloud

0 packages can be updated.
0 updates are security updates.


Last login: Wed Apr 27 07:38:52 2016 from 192.168.120.1
user@myvnf:~$
```

第 5 章 OpenStack Tacker による NFV オーケストレーション

　今回しようしている VNF のファイアーウォール機能には iptables を使用しています。net0 から net1 への転送を拒否するように設定を変更します。

```
user@myvnf:~$ sudo iptables -A FORWARD -i eth1 -o eth2 -j REJECT
user@myvnf:~$ sudo iptables -L
sudo: unable to resolve host myvnf
Chain INPUT (policy ACCEPT)
target     prot opt source               destination

Chain FORWARD (policy ACCEPT)
target     prot opt source               destination
REJECT     all  --  anywhere             anywhere             reject-with
icmp-port-unreachable

Chain OUTPUT (policy ACCEPT)
target     prot opt source               destination
```

net0-vm から net1-vm への疎通を確認してみます。

```
$ ping 10.10.1.4 -c 5
PING 10.10.1.4 (10.10.1.4): 56 data bytes

--- 10.10.1.4 ping statistics ---
5 packets transmitted, 0 packets received, 100% packet loss
```

　VNF の設定を変更したことにより、net0-vm から net1-vm への ping 疎通ができなくなったことが確認できました。

　再度、通信を許可するには以下の様に設定します。

```
~/devstack$ sudo iptables -D FORWARD -i eth1 -o eth2 -j REJECT
```

　本章は OpenStack で NFV を実現する Tacker をご紹介しました。ぜひ OpenStack Tacker を使った NFV の構築にチャレンジしてみて下さい。OpenDaylight と連携するための実装も進められているので、ドキュメントも充実していき、OpenDaylight ユーザにとってもますます注目すべきコンポーネントとなっていくでしょう。

第6章 OpenDaylightでクラスタを組んでみよう

　前章までは、OpenDaylight（以下 ODL）の構築と基本動作確認を行ってきました。OpenStackとの連携もできているので、使ってみようかなと思う方もおられるかと思います。実際に、システム構成を考えるときに SPOF（単一障害点）を作らないように設計するのは基本かと思いますが、OpenStack は HA 構成が組めるのに、ODL が SPOF になってしまっては元も子もありません。そこで本章は、ODL のクラスタを構築してみます。

6.1　OpenDaylight のデータストアとクラスタ

　ODL のクラスタは MD-SAL Clustering 機能として実装されています。ODL の分散データストアを理解するためにはデータストアがどのような構造になっているのかを理解している必要があります。

データストア

　OpenDaylight は MD-SAL という抽象化レイヤー内に独自のデータストアを持っています。ODL を利用する各アプリケーションはこのデータストアにアクセスすることで必要な情報を保存したり、取り出したりできます。データストアの実体は LevelDB(Key-Value Storage) か Snapshot となりますが、論理的には以下の図のようなデータツリー構造をしています（図6.1）。

シャード

　クラスタを構成した際、データストアに保存されたデータはクラスタ内の多数のノードへシャード単位でレプリケーションされます。シャードとはデータストアのあるひとまとめにし

第 6 章　OpenDaylight でクラスタを組んでみよう

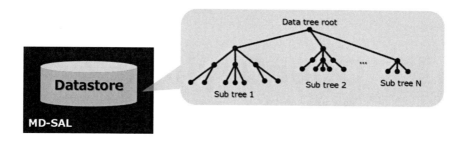

図 6.1　Data tree

たサブツリーを指します。シャード毎にどのノードへレプリケーションするかを設定できます（図 6.2）。

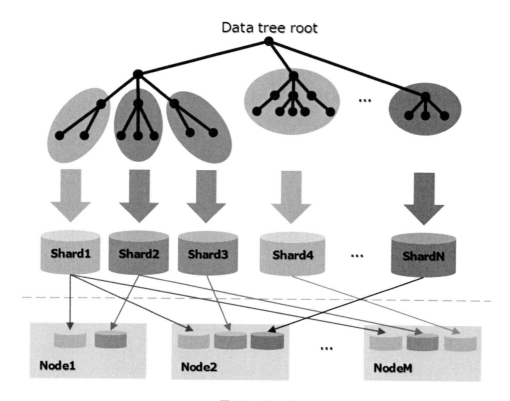

図 6.2　Shard

6.2 Raft

ODL のクラスタでは Raft というコンセンサスアルゴリズムを採用しています。Raft では各ノードは以下の 3 つの状態のいずれかを持っています。

- Leader
  - Leader はクラスタ内で 1 ノードのみ選出され、ODL アプリケーションや Follower からの全ての要求に応答し、Follower への要求も行います。
- Follower
  - Follower は Leader か Candidate のノードからの要求に応答します。自ら要求を行うことはありません。
- Candidate
  - Candidate はノードの起動時や Leader がダウンした時に、新しい Leader を選出する状態を示しています。

Raft ではクラスタ内の過半数以上のノードが稼働していれば全体のフォールトトレラント性が保たれる仕組みになっています。もし、何らかの障害により過半数以上のノードがダウンした場合、Leader を選出できずクラスタは Suspend 状態となります。Raft では全てのノードがリクエストを受けることができます。Follower が要求を受けるとその要求は Leader に転送されます。したがって、OpenStack からのリクエストが Leader ではないノードに届いても問題はありません。実際に、今回解説する構成では HAProxy を使ってロードバランシングしていますので、Follower にもリクエストが飛ぶことになります。

6.3 環境・バージョンについて

クラスタの物理構成

今回は 3 ノードの ODL クラスタを構築してみます。これまでと同様に OpenStack と連携させた環境を構築します。OpenStack 用の 1 ノードと ODL クラスタ用の 3 ノードの全てで Ubuntu を使用します。ODL へのリクエストを ODL クラスタへ分散するために OpenStack のノードに HAProxy も使用します。それぞれのパッケージで使用するバージョンは以下の表の通りです。

第 6 章　OpenDaylight でクラスタを組んでみよう

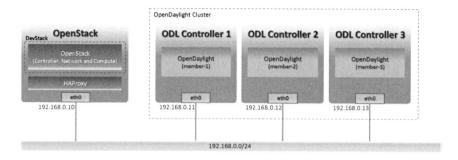

図 6.3　env

パッケージ	バージョン
Ubuntu	14.04 LTS
OpenDaylight	Beryllium SR2
OpenStack	Mitaka
HAProxy	1.4.24

シャードの構成

今回の構成では以下の図のように、Inventory サブツリー配下のシャードと network-topology サブツリー配下のシャード、それら以外のサブツリーである default、toaster のシャードを全てのノードにレプリケーションする設定を行います。

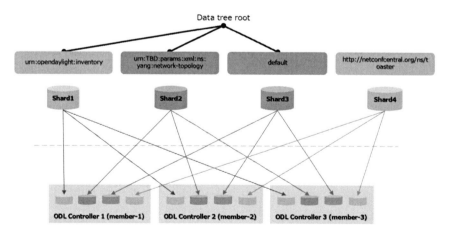

図 6.4　Data Tree

6.4　構築してみよう

全ノード共通の設定

それでは早速構築を始めていきます。先に、全ノードで共通の設定を実施しておきます。構築を簡単にするために、ファイアーウォールは無効化しておきます。

```
~$ sudo ufw disable
```

続いて、ODL の構築に入ります。

ODL の構築

JDK のインストール

ODL は Java 言語によって実装されています。そのため Java の実行環境が必要になりますので JDK(Java Development Kit) をインストールします。今回インストールする Beryllium SR2 では JDK 8 が必要です。

```
~$ sudo add-apt-repository ppa:openjdk-r/ppa
~$ sudo apt-get update
~$ sudo apt-get install openjdk-8-jre
```

OpenDaylight のダウンロード

ODL のパッケージを Nexus リポジトリからダウンロードして展開します。

```
~$ wget https://nexus.opendaylight.org/content/repositories/opendaylight.release/org/opendaylight/integration/distribution-karaf/0.4.2-Beryllium-SR2/distribution-karaf-0.4.2-Beryllium-SR2.tar.gz
~$ tar zxvf distribution-karaf-0.4.2-Beryllium-SR2.tar.gz
~$ cd distribution-karaf-0.4.2-Beryllium-SR2/
```

OpenDaylight の起動

ODL を以下の start スクリプトを実行して起動します。

第 6 章　OpenDaylight でクラスタを組んでみよう

```
~/distribution-karaf-0.4.2-Beryllium-SR2$ ./bin/start   # OpenDaylight を起動する
```

必要な機能のインストール

　ODL のコンソールにログインして、必要な機能をインストールしていきます。インストールするのは、OpenStack 連携に必要な「odl-ovsdb-openstack」と MD-SAL のクラスタリングに必要な「odl-mdsal-clustering」になります。

```
~/distribution-karaf-0.4.2-Beryllium-SR2$ ./bin/client -u karaf   # OpenDaylight
Karaf Shell にログイン
   :
opendaylight-user@root>feature:install odl-ovsdb-openstack
opendaylight-user@root>feature:install odl-mdsal-clustering
opendaylight-user@root>logout
```

　ここで、一度 ODL を停止します。

```
~/distribution-karaf-0.4.2-Beryllium-SR2$ ./bin/stop   # OpenDaylight を停止する
```

　ここまでの手順を 3 台の ODL ノード全てで実施してください。

クラスタリングの設定

　続いて、クラスタリングに必要な設定をしていきます。ODL クラスタリングでは akka.conf と module-shards.conf の 2 つのファイルを設定する必要があります。今回使用する設定ファイルは全て GitHub からダウンロードできます。

　まずは ODL Controller 1 のノードで設定を行います。以下のディレクトリに移動し、akka.conf をダウンロードします。

```
~/distribution-karaf-0.4.2-Beryllium-SR2$ cd configuration/initial
~/distribution-karaf-0.4.2-Beryllium-SR2/configuration/initial$ wget -O akka.conf
https://raw.githubusercontent.com/YujiAzama/opendaylight-openstack-integration/master
/clustering/akka.conf.member-1
```

　ODL Controller 2 と 3 の akka.conf は以下の URL からダウンロードして使用してください。

- ODL Controller 2
    - https://raw.githubusercontent.com/YujiAzama/opendaylight-openstack-integration/master/clustering/akka.conf.member-2
- ODL Controller 3

- https://raw.githubusercontent.com/YujiAzama/opendaylight-openstack-integration/master/clustering/akka.conf.member-3

ダウンロードしたファイルを開き、hostname や seed-nodes の IP アドレスをお使いの環境に合わせて編集してください。roles にはクラスタ内でのメンバー識別子が指定されています。それぞれのメンバーはクラスタ内での識別子を持っている必要があり、そのクラスタ内で一意でなければなりません。ここでは以下のように値を設定しています。

ノード	メンバー識別子
ODL Controller 1	member-1
ODL Controller 2	member-2
ODL Controller 3	member-3

```
odl-cluster-data {
  akka {
    remote {
      netty.tcp {
        hostname = "192.168.0.11"   # 自ノードの IP アドレスへ変更する
        port = 2550
      }
    }

    cluster {
      seed-nodes = ["akka.tcp://opendaylight-cluster-data@192.168.0.11:2550",   # member-1 の IP アドレスを指定する
                    "akka.tcp://opendaylight-cluster-data@192.168.0.12:2550",   # member-2 の IP アドレスを指定する
                    "akka.tcp://opendaylight-cluster-data@192.168.0.13:2550"]   # member-3 の IP アドレスを指定する

      roles = [
        "member-1"   # クラスタ内でのメンバー識別子を指定する
      ]

    }
    :   # 省略
}

odl-cluster-rpc {
  :   # 省略
    remote {
      log-remote-lifecycle-events = off
      netty.tcp {
        hostname = "192.168.0.11"   # 自ノードの IP アドレスへ変更する
        port = 2551
        maximum-frame-size = 419430400
        send-buffer-size = 52428800
        receive-buffer-size = 52428800
```

第 6 章　OpenDaylight でクラスタを組んでみよう

```
    }
  }
  cluster {
    seed-nodes = ["akka.tcp://odl-cluster-rpc@192.168.0.11:2551",  # member-1 の IP
アドレスを指定する
                  "akka.tcp://odl-cluster-rpc@192.168.0.12:2551",  # member-2 の IP
アドレスを指定する
                  "akka.tcp://odl-cluster-rpc@192.168.0.13:2551"]  # member-3 の IP
アドレスを指定する
    auto-down-unreachable-after = 300s
  }
 }
}
```

次に、module-shards.conf をダウンロードします。

```
~/distribution-karaf-0.4.2-Beryllium-SR2/configuration/initial$ wget -O
module-shards.conf
https://raw.githubusercontent.com/YujiAzama/opendaylight-openstack-integration/master
/clustering/module-shards.conf
```

module-shards.conf ではクラスタのメンバー (ODL のノード) とレプリカの設定をします。レプリカのリストの順番に優先順位が設定されます。

```
module-shards = [
    {
        name = "default"
        shards = [
            {
                name="default"
                replicas = [
                    "member-1",
                    "member-2",
                    "member-3"
                ]
            }
        ]
    },
    {
        name = "topology"
        shards = [
            {
                name="topology"
                replicas = [
                    "member-1",
                    "member-2",
                    "member-3"
                ]
            }
        ]
```

```
        },
        {
            name = "inventory"
            shards = [
                {
                    name="inventory"
                    replicas = [
                        "member-1",
                        "member-2",
                        "member-3"
                    ]
                }
            ]
        },
        {
            name = "toaster"
            shards = [
                {
                    name="toaster"
                    replicas = [
                        "member-1",
                        "member-2",
                        "member-3"
                    ]
                }
            ]
        }
]
```

ODL Controller 2 と ODL Controller 3 も同様に設定をします。最後に、全ての ODL を起動してください。

```
~/distribution-karaf-0.4.2-Beryllium-SR2/configuration/initial$ cd
~/distribution-karaf-0.4.2-Beryllium-SR2$
~/distribution-karaf-0.4.2-Beryllium-SR2$ ./bin/start   # OpenDaylight を起動する
```

以上で、ODL の設定は完了です。

HAProxy の設定

今回は ODL がクラスタ構成となっていますので、Neutron ML2 ドライバからのリクエストをクラスタの各メンバーへ分散するために HAProxy を使用してみます。OpenStack のノードでインストールと設定を行っていきます。

```
~$ sudo apt-get install haproxy
```

以下の設定ファイルを開いてください。

```
~$ sudo vim /etc/default/haproxy
```

ENABLED が 0 になっていれば 1 に変更してください。

```
ENABLED=1
```

続いて、設定ファイルをダウンロードします。

```
~$ sudo wget -O /etc/haproxy/haproxy.cfg
https://raw.githubusercontent.com/YujiAzama/opendaylight-openstack-integration/master
/clustering/haproxy.cfg
```

ダウンロードした haproxy.cfg を開き backend_servers に指定されているクラスタのメンバーの IP アドレスへ変更してください。HAProxy の設定は非常に単純で、8181 番ポートから入ってきた http パケットをバックエンドの ODL へラウンドロビンで分散するように設定しています。

```
global
    log /dev/log    local0
    log /dev/log    local1 notice
    chroot /var/lib/haproxy
    user haproxy
    group haproxy
    daemon

defaults
    log global
    mode    http
    option  httplog
    option  dontlognull
        contimeout 5000
        clitimeout 50000
        srvtimeout 50000
    errorfile 400 /etc/haproxy/errors/400.http
    errorfile 403 /etc/haproxy/errors/403.http
    errorfile 408 /etc/haproxy/errors/408.http
    errorfile 500 /etc/haproxy/errors/500.http
    errorfile 502 /etc/haproxy/errors/502.http
    errorfile 503 /etc/haproxy/errors/503.http
    errorfile 504 /etc/haproxy/errors/504.http

frontend http-in
        bind *:8181
        default_backend backend_servers
        option            forwardfor

backend backend_servers
        balance roundrobin
```

```
        server odl1 192.168.0.11:8181 check    # member-1のIPアドレスへ変更
        server odl2 192.168.0.12:8181 check    # member-2のIPアドレスへ変更
        server odl3 192.168.0.13:8181 check    # member-3のIPアドレスへ変更
```

最後にHAProxyを再起動します。

```
~$ sudo service haproxy restart
```

以上で、HAProxyの設定は完了です。

コントローラノードの構築

続いて、OpenStackの構築を行います。DevStackをダウンロードしてきます。

```
~$ sudo apt-get install git
~$ git clone https://github.com/openstack-dev/devstack.git --branch stable/mitaka
```

以下のURLからlocal.confのサンプルをダウンロードしてください。

```
~$ cd devstack/
~/devstack$ wget
https://raw.githubusercontent.com/YujiAzama/opendaylight-openstack-integration/master
/clustering/local.conf
```

ダウンロードしたlocal.confを開くと以下の様になっています。まず、お使いの環境に合わせてHOST_IPをコントローラノードのIPアドレスに変更してください。

```
~/devstack$ vim local.conf
  :
# IP Details
HOST_IP=192.168.0.10
SERVICE_HOST=$HOST_IP
  :
# Neutron
  :
BRANCH_NAME=stable/beryllium
enable_plugin networking-odl http://git.openstack.org/openstack/networking-odl
${BRANCH_NAME}

# OpenDaylight Details
ODL_MODE=externalodl
ODL_PORT=8181
ODL_OVS_MANAGERS=192.168.0.11,192.168.0.12,192.168.0.13

[[post-config|/etc/neutron/plugins/ml2/ml2_conf.ini]]
[ml2_odl]
password=admin
username=admin
```

```
url="http://127.0.0.1:${ODL_PORT}/controller/nb/v2/neutron"
```

　ODL_OVS_MANAGERS には、OpenStack の各 OVS が接続する先の ODL の IP アドレスを指定します。今回は 3 台の ODL がいるので、3 台分の IP アドレスを指定します。本来、ml2_conf.ini の ml2_odl セクションには ODL コントローラの IP アドレスを url として指定しますが、今回の ODL はクラスタ構成なので、HAProxy が動作しているローカルホストを指定しています。

　後は、stack.sh スクリプトを実行するだけです。以下の様なメッセージが出力されれば終了です。

```
~/devstack$ ./stack.sh
    :
This is your host IP address: 192.168.0.10
This is your host IPv6 address: ::1
Horizon is now available at http://192.168.0.10/dashboard
Keystone is serving at http://192.168.0.10:5000/
The default users are: admin and demo
The password: password
2016-07-22 06:47:58.133 | stack.sh completed in 1130 seconds.
```

　以上で全ての構築は完了です。

6.5　動作確認

レプリケーションされているか

　機能的には第 2 章、第 3 章で構築した環境と変わりません。ここでは、ODL のデータストアが全てのメンバーにレプリケーションできているか確認してみましょう。

　ODL にデータを登録するために OpenStack からネットワークとインスタンスを作成します。コントローラノードで以下のコマンドを実行します。

```
~/devstack$ source openrc demo demo      # 認証用の環境変数を設定 (demo ユーザー、demo テナント)
~/devstack$ neutron net-create net01     # ネットワークを作成
Created a new network:
+---------------------------+--------------------------------------+
| Field                     | Value                                |
+---------------------------+--------------------------------------+
| admin_state_up            | True                                 |
| availability_zone_hints   |                                      |
| availability_zones        |                                      |
| created_at                | 2016-07-26T07:43:08                  |
```

```
| description             |                                        |
| id                      | b94883ed-34ad-4a09-ac58-e459b6b77582   |
| ipv4_address_scope      |                                        |
| ipv6_address_scope      |                                        |
| mtu                     | 1450                                   |
| name                    | net01                                  |
| port_security_enabled   | True                                   |
| router:external         | False                                  |
| shared                  | False                                  |
| status                  | ACTIVE                                 |
| subnets                 |                                        |
| tags                    |                                        |
| tenant_id               | aaa0ecf59218482d9e55b75a50f22a5c       |
| updated_at              | 2016-07-26T07:43:08                    |
+-------------------------+----------------------------------------+
```

net01 というネットワークが作成され、ODL のデータストアに情報が書き込まれました。さっそくデータストアの情報を確認してみましょう。第 4 章で紹介した RESTCONF を使ってアクセスします。まずは member-1 にアクセスして neutron の情報を取得します。以下のようなデータが取得できます。

```
~$ curl -u admin:admin http://192.168.0.11:8181/restconf/config/neutron:neutron/ | python -m json.tool
{
    "neutron": {
        "networks": {
            "network": [
                {
                    "admin-state-up": true,
                    "name": "net01",
                    "neutron-L3-ext:external": false,
                    "neutron-provider-ext:network-type": "neutron-networks:network-type-vxlan",
                    "neutron-provider-ext:segmentation-id": "1048",
                    "shared": false,
                    "status": "ACTIVE",
                    "tenant-id": "aaa0ecf5-9218-482d-9e55-b75a50f22a5c",
                    "uuid": "b94883ed-34ad-4a09-ac58-e459b6b77582"
                }
            ]
        },
        :
    }
}
```

データ量が多いので一部省略していますが、先ほど作成したネットワークの情報が取得できます。member-2 と member-3 に対しても同様に RESTCONF でアクセスしてみてください。同じデータが取得できていればレプリケーションできており、クラスタが正常に動作しています。

障害時の動作

それではクラスタの状態を確認してみます。Java アプリケーションの統計情報を取得したりモニタリングや管理などを行う時に欠かせない JMX ですが、ODL では JMX の情報を REST API で提供するための Jolokia を利用できます。まず、Jolokia をインストールします。

```
~/distribution-karaf-0.4.2-Beryllium-SR2$ ./bin/client -u karaf
client: JAVA_HOME not set; results may vary
Logging in as karaf
387 [sshd-SshClient[ed17bee]-nio2-thread-2] WARN org.apache.sshd.client.keyverifier.AcceptAllServerKeyVerifier - Server at [/0.0.0.0:8101, DSA, 75:e8:b1:02:c3:ac:e8:6d:f8:8e:fb:99:a8:b2:1a:8e] presented unverified {} key: {}

    _____                       _____                      .__  .__       .__     __
    \_____  \   _____   ____    _____ \ _____  ___.___.|  | |__| ____ |  |___/  |_
     /   |   \ /  ____\_/ __ \  /    \|    |  \\__  \<   |  ||  | |  |/ ___\|  |  \   __\
    /    |    \\  |__>  >  ___/ |   |  \    `   \/ __ \\___  ||  |_|  / /_/  >   Y  \  |
    _____  / \___  /  \___  >|___|  /_____  (____  / ____||____/__\___  /|___|  /__|
            \/|__|     \/         \/       \/        \/\/     \/      /_____/       \/

Hit '<tab>' for a list of available commands
and '[cmd] --help' for help on a specific command.
Hit '<ctrl-d>' or type 'system:shutdown' or 'logout' to shutdown OpenDaylight.

opendaylight-user@root>bundle:install -s mvn:org.jolokia/jolokia-osgi/1.3.1
Bundle ID: 286
opendaylight-user@root>logout
```

Jolokia のインストールを全ての member に対して行います。

まず member-1 の状態を取得してみます。

```
$ curl -s http://192.168.0.11:8181/jolokia/read/org.opendaylight.controller:Category=Shards,name=member-1-shard-inventory-config,type=DistributedConfigDatastore | python -m json.tool
{
    "request": {
        "mbean": "org.opendaylight.controller:Category=Shards,name=member-1-shard-inventory-config,
                  type=DistributedConfigDatastore",
        "type": "read"
    },
    "status": 200,
    "timestamp": 1465515563,
    "value": {
        "AbortTransactionsCount": 0,
        "CommitIndex": 13,
        "CommittedTransactionsCount": 0,
        "CurrentTerm": 5,
```

```
        "FailedReadTransactionsCount": 0,
        "FailedTransactionsCount": 0,
        "FollowerInfo": [
            {
                "active": true,
                "id": "member-2-shard-inventory-config",
                "matchIndex": 13,
                "nextIndex": 14,
                "timeSinceLastActivity": "00:00:00.012"
            },
            {
                "active": true,
                "id": "member-3-shard-inventory-config",
                "matchIndex": 13,
                "nextIndex": 14,
                "timeSinceLastActivity": "00:00:00.015"
            }
        ],
        "FollowerInitialSyncStatus": true,
        "InMemoryJournalDataSize": 1338,
        "InMemoryJournalLogSize": 1,
        "LastApplied": 13,
        "LastCommittedTransactionTime": "1970-01-01 09:00:00.000",
        "LastIndex": 13,
        "LastLeadershipChangeTime": "2016-06-10 08:37:53.140",
        "LastLogIndex": 13,
        "LastLogTerm": 4,
        "LastTerm": 4,
        "Leader": "member-1-shard-inventory-config",
        "LeadershipChangeCount": 3,
        "PeerAddresses": "member-2-shard-inventory-config:
akka.tcp://opendaylight-cluster-data@192.168.0.12:2550/user/shardmanager-config
                            /member-2-shard-inventory-config,
member-3-shard-inventory-config:
akka.tcp://opendaylight-cluster-data@192.168.0.13:2550/user/shardmanager-config
                            /member-3-shard-inventory-config",
        "PendingTxCommitQueueSize": 0,
        "RaftState": "Leader",
        "ReadOnlyTransactionCount": 0,
        "ReadWriteTransactionCount": 0,
        "ReplicatedToAllIndex": 12,
        "ShardName": "member-1-shard-inventory-config",
        "SnapshotCaptureInitiated": false,
        "SnapshotIndex": 12,
        "SnapshotTerm": 4,
        "StatRetrievalError": null,
        "StatRetrievalTime": "1.676 ms",
        "TxCohortCacheSize": 0,
        "VotedFor": "member-1-shard-inventory-config",
        "WriteOnlyTransactionCount": 0
    }
}
```

取得した結果、"RaftState": "Leader" となっていることが確認できます。続いて、member-2 と member-3 でも確認してみてください。"RaftState": "Follower" となっていることが確認できます。

member-1 を落としてみる

クラスタが正常に動作していることを確認するために、Leader となっている member-1 を停止して member-2 に Leader が切り替わることを確認します。

```
~/distribution-karaf-0.4.2-Beryllium-SR2$ ./bin/stop
```

これで member-1 からのレスポンスは返ってこなくなります。この時の member-2 の状態を確認してみます。

```
$ curl -s
http://192.168.0.12:8181/jolokia/read/org.opendaylight.controller:Category=Shards,
name=member-2-shard-inventory-config,type=DistributedConfigDatastore | python -m
json.tool
{
    "request": {
        "mbean":
"org.opendaylight.controller:Category=Shards,name=member-2-shard-inventory-config,
                        type=DistributedConfigDatastore",
        "type": "read"
    },
    "status": 200,
    "timestamp": 1465449951,
    "value": {
        "AbortTransactionsCount": 0,
        "CommitIndex": 13,
        "CommittedTransactionsCount": 0,
        "CurrentTerm": 7,
        "FailedReadTransactionsCount": 0,
        "FailedTransactionsCount": 0,
        "FollowerInfo": [
            {
                "active": false,
                "id": "member-1-shard-inventory-config",
                "matchIndex": -1,
                "nextIndex": 13,
                "timeSinceLastActivity": "00:00:00.000"
            },
            {
                "active": false,
                "id": "member-3-shard-inventory-config",
                "matchIndex": -1,
                "nextIndex": 13,
                "timeSinceLastActivity": "00:00:00.000"
```

```
            }
        ],
        "FollowerInitialSyncStatus": true,
        "InMemoryJournalDataSize": 1338,
        "InMemoryJournalLogSize": 1,
        "LastApplied": 13,
        "LastCommittedTransactionTime": "1970-01-01 09:00:00.000",
        "LastIndex": 13,
        "LastLeadershipChangeTime": "2016-06-09 14:25:51.866",
        "LastLogIndex": 13,
        "LastLogTerm": 4,
        "LastTerm": 4,
        "Leader": "member-2-shard-inventory-config",
        "LeadershipChangeCount": 4,
        "PeerAddresses": "member-1-shard-inventory-config:
akka.tcp://opendaylight-cluster-data@192.168.0.11:2550/user/shardmanager-config
                          /member-1-shard-inventory-config,
member-3-shard-inventory-config:
akka.tcp://opendaylight-cluster-data@192.168.0.13:2550/user/shardmanager-config
                          /member-3-shard-inventory-config",
        "PendingTxCommitQueueSize": 0,
        "RaftState": "Leader",
        "ReadOnlyTransactionCount": 1,
        "ReadWriteTransactionCount": 0,
        "ReplicatedToAllIndex": -1,
        "ShardName": "member-2-shard-inventory-config",
        "SnapshotCaptureInitiated": false,
        "SnapshotIndex": 12,
        "SnapshotTerm": 4,
        "StatRetrievalError": null,
        "StatRetrievalTime": "19.88 ms",
        "TxCohortCacheSize": 0,
        "VotedFor": "member-2-shard-inventory-config",
        "WriteOnlyTransactionCount": 0
    }
}
```

member-1 が停止したことによって Leader が member-2 に切り替わりました。この状態でも OpenStack のネットワークが作れるか確認します。

```
~/devstack$ neutron net-create net02
Created a new network:
+---------------------------+--------------------------------------+
| Field                     | Value                                |
+---------------------------+--------------------------------------+
| admin_state_up            | True                                 |
| availability_zone_hints   |                                      |
| availability_zones        |                                      |
| created_at                | 2016-07-26T08:01:35                  |
| description               |                                      |
| id                        | f9867311-98c3-4618-be98-6ce75ff8aa23 |
```

第 6 章　OpenDaylight でクラスタを組んでみよう

```
| ipv4_address_scope    |                                      |
| ipv6_address_scope    |                                      |
| mtu                   | 1450                                 |
| name                  | net02                                |
| port_security_enabled | True                                 |
| router:external       | False                                |
| shared                | False                                |
| status                | ACTIVE                               |
| subnets               |                                      |
| tags                  |                                      |
| tenant_id             | aaa0ecf59218482d9e55b75a50f22a5c     |
| updated_at            | 2016-07-26T08:01:35                  |
+-----------------------+--------------------------------------+
```

作成できました。member-1 を起動して net02 がレプリケーションされることを確認しましょう。

```
~/distribution-karaf-0.4.2-Beryllium-SR2$ ./bin/start
```

```
~$ curl -u admin:admin http://192.168.0.11:8181/restconf/config/neutron:neutron/
{
    "neutron": {
        "networks": {
            "network": [
                {
                    "admin-state-up": true,
                    "name": "net01",
                    "neutron-L3-ext:external": false,
                    "neutron-provider-ext:network-type":
"neutron-networks:network-type-vxlan",
                    "neutron-provider-ext:segmentation-id": "1048",
                    "shared": false,
                    "status": "ACTIVE",
                    "tenant-id": "aaa0ecf5-9218-482d-9e55-b75a50f22a5c",
                    "uuid": "b94883ed-34ad-4a09-ac58-e459b6b77582"
                },
                {
                    "admin-state-up": true,
                    "name": "net02",
                    "neutron-L3-ext:external": false,
                    "neutron-provider-ext:network-type":
"neutron-networks:network-type-vxlan",
                    "neutron-provider-ext:segmentation-id": "1069",
                    "shared": false,
                    "status": "ACTIVE",
                    "tenant-id": "aaa0ecf5-9218-482d-9e55-b75a50f22a5c",
                    "uuid": "f9867311-98c3-4618-be98-6ce75ff8aa23"
                }
            ]
        },
        :
```

```
    }
}
```

　再起動した member-1 にも正しくレプリケーションされていることが確認できました。以上で、動作確認は終了です。

　本章は、第 2 章と第 3 章で解説した OpenStack 連携や第 4 章で解説した RESTCONF などを取り入れて OpenDaylight のクラスタリングを解説しました。これらのように、OpenDaylight はクライアントアプリケーションにとって汎用性が高く、HA も考慮されている設計になってきています。

付録A　Service Function Chaining

A.1　SFCとは何か？

　企業やキャリアなどにおいてネットワークを設計する際、ファイアーウォールやロードバランサー、IDS、NATなど、さまざまなアプライアンスを組み合わせると思います。これらのアプライアンスはNFVの世界ではネットワークファンクション (Network Function) と呼んだり、1つのネットワークサービスとみなしてサービスファンクション (Service Function) と呼んだりします。このSFの組み合わせ、つまり、「パケットが通過するサービスの順序」をよりプログラマブルに定義できるようにしたものをサービスファンクションチェイニング (Service Function Chaining, SFC) と呼びます。SFCはIETFのRFC7665において標準化が行われており、OpenDaylightでもプラグインを利用することによってSFCを実現することができます。

A.2　用語の解説

　SFCでは普段使わないネットワークの用語が多く出てきます。基本的な用語を理解していないとSFCの概要が理解できないので解説していきます。

ネットワークサービス (Network Service, NS)

　ネットワークサービスとは、オペレータから提供されるサービスのことで、1つ以上のサービスファンクションによって構成されるもののことを指します。

付録 A　Service Function Chaining

クラシフィケーション (Classification)

クラシフィケーションとは、サービスファンクションパスを決定するためのポリシーです。クラシフィケーションで定義されたポリシーの条件から、入ってきたパケットの転送先のサービスファンクションパスが決定され、適切なサービスファンクションに転送されます。

クラシファイア (Classifier)

クラシファイアとは、クラシフィケーションとして実際に動作するエレメントです。

サービスファンクションチェイン (Service Function Chain, SFC)

サービスファンクションチェインとは、抽象的なサービスファンクションの順序を定義したものです。抽象化されたものなので、定義する際にサービスファンクションの物理的な配置は気にしません。ファイアーウォール→ DPI →ロードバランサーのように、パケットを流したいサービスファンクションの順番を定義します。

サービスファンクション (Service Function, SF)

サービスファンクションは、プロトコルスタックの様々なレイヤーにおいて、受信したパケットに基づいて特定の処理を行う機能を持っています。先にも上げた、ファイアーウォールやロードバランサー、IDS、NAT、WAN アクセラレーションなどが SF にあたります。SF は論理的なコンポーネントですが、仮想アプライアンスなどの VM や物理的なネットワークのエレメントとして実現されます。

サービスファンクションフォワーダー (Service Function Forwarder, SFF)

サービスファンクションフォワーダーは、カプセル化されて運ばれてきた SFC のヘッダー情報にしたがって、1 つ以上のサービスファンクションへトラフィックを転送する機能を持っています。

メタデータ (Metadata)

メタデータは、クラシファイアおよびサービスファンクション間でコンテキスト情報の交換を行う機能を持っています。

サービスファンクションパス (Service Function Path, SFP)

サービスファンクションパスとは、サービスファンクションチェインの定義から、実際にパケットが通過するサービスファンクションインスタンスの順番を定義したものです。

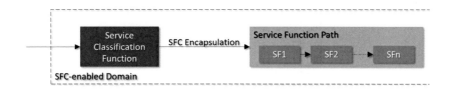

図 A.1　Service Function Path

SFC カプセル化 (SFC Encapsulation)

SFC カプセル化では、SFP 識別子を持っており、サービスファンクションフォワーダーやサービスファンクションチェインを認識できるサービスファンクションに利用されます。SFP 識別子を持っていることによってサービスファンクションパスの特定を可能にします。SFC カプセル化は SFP 識別子に加えて、データプレーンのコンテキスト情報を含むメタデータを運ぶ役割も持っています。SFC カプセル化の方式には Network Service Header(NSH) や Service Chain Header(SCH) があります。

レンダードサービスパス (Rendered Service Path, RSP)

レンダードサービスパスとは、実際にパケットが流れるサービスファンクションフォワーダーとサービスファンクションの実際の順序です。サービスファンクションパスよりもさらに具体的なパケットの流れる順序になります。

SFC プロキシ (SFC Proxy)

SFC プロキシは SFC を認識できないサービスファンクションに代わって SFC カプセル化の挿入と削除を行います。これによって、SFC に対応していない仮想アプライアンスなどもサービスファンクションパス内で使用することができるようになります。

付録 A　Service Function Chaining

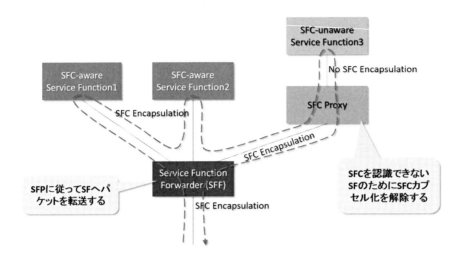

図 A.2　Service Function Proxy

A.3　OpenDaylight の SFC 機能

　OpenDaylight では RFC7665 の標準化に基づいた SFC の実装を行っています。ODL の SFC はプラグインで実装されており、OVSDB を用いる場合、odl-ovsdb-sfc などをインストールすることによって利用することができます。実際に利用する際には、サービスファンクションフォワーダーを定義し、そこにサービスファンクションを登録していきます。サービスファンクションは管理用 IP アドレスや URI、REST エンドポイントの設定が行えるので、仮想と物理を問わずさまざまなアプライアンスが登録できます。さらに、SFC カプセル化に対応していないサービスファンクションのために SFC プロキシも利用できます。パケットの入り口となるデータプレーンのエンドポイントの IP アドレスもサービスファンクションフォワーダーで設定します。パケットを別のサービスファンクションフォワーダーへ転送したい場合には、VxLAN などのトンネリングが利用できます（図 1.3）。

　事前に登録したサービスファンクションのリストからサービスファンクションチェインを定義していきます。DLUX の GUI から行う場合にはドラッグアンドドロップで行います（図 1.4）。

　サービスファンクションチェインの定義から、実際のパケットが流れる経路になるサービスファンクションパスやレンダードサービスパスをデプロイできます。これらの設定は全て API から行うことができるので、OpenStack などの ODL アプリケーションはその API を利用することになります。

付録 A　Service Function Chaining

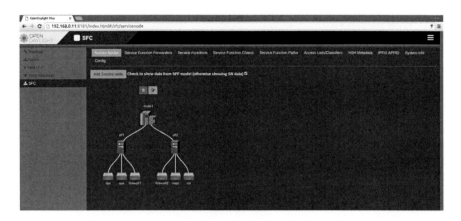

図 A.3　Add Service Function

図 A.4　Service Function Chaining

A.4　OpenStackにおけるSFCへの取り組み

　OpenStackにおいてもnetworking-sfcというNeutronのサブプロジェクトで、OpenStack上でSFCを実現するための議論が活発に行われています。networking-sfcはNeutronのプラグインとして実装されています。networking-sfcプラグインはCLIからクライアントコマンドによるオペレーションができるように開発が進められています。さらに、第5章の中で紹介したOpenStack Tackerのプロジェクトでは、networking-sfcのAPIを呼び出すためのドライバが開発されており、これらが利用できるようになると、TOSCAファイルにSFCの構成を定義しておきTackerに読み込ませ、networking-sfcやnetwokring-odlプラグインを使うことによって、ODLのSFC機能を使ったサービスファンクションチェインのプロビジョニングができるようになります。networking-sfcではOpenDaylightの他にもOVSやONOS(The Open Network

付録 A　Service Function Chaining

図 A.5　Rendered Service Path

Operating System) などの SFC もサポートする計画になっています（図 1.6）。

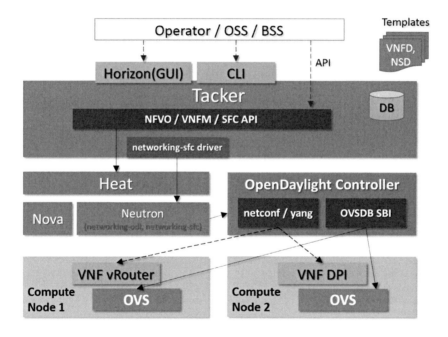

図 A.6　Tacher SFC Overview

これまで紹介してきたように、OpenDaylight や OpenStack などの技術の連携によって、NFV の要件が実現されようとしています。これらの技術に興味のあった方、本書を読んで興味の出た方は、ぜひ一度 OpenDaylight や OpenStack にチャレンジしてみてください。

●著者紹介

倉橋　良
NEC ソリューションイノベータ株式会社
沖縄オープンラボラトリ研究メンバー。元々は Web のソフトウェアエンジニアとして映像配信のシステムやクラウドポータルの開発を行ってきた。現在は沖縄オープンラボラトリの研究メンバーとして OpenStack や OpenDaylight などのクラウド・SDN 領域における開発・研究活動を日々行っている。

鳥居　隆史
NEC OSS 推進センター
沖縄オープンラボラトリ技術主査。OpenStack が公開された当初から、ネットワークに関係する部分の開発や検証を行ってきている。また、日本 OpenStack ユーザー会のボードメンバーとしてイベントや発表多数。さらに、沖縄オープンラボのメンバーとして OpenStack と SDN の融合分野における先端技術の開発・検証を主導。

安座間　勇二
NEC ソリューションイノベータ株式会社
1991 年、沖縄県生まれ。2014 年の入社以来、ソフトウェアエンジニアとして OpenStack での SFC(Service Function Chaining) や OpenDaylight を使った SDN/NFV を中心とした開発や検証などに携わる。OpenStack Neutron を中心にコントリビューションしており、OpenStack Summit Tokyo 2015 において SFC の R&D について発表。

高橋 信行
株式会社アインザ
主に通信主体の業務に携わり、元々は、企業向けネットワークサービスを提供 (主に VPN) する業務に従事し、L2、L3 通信機器の設定 (主に Cisco systems、Alcatel lucent、Alaxala Networks、NEC)、ネットワーク機器の検証などを行ってきた。2015 年 4 月から、ネットワークエンジニアとしての経験より沖縄オープンラボラトリの活動に参加し、OpenStack での SFC(Service Function Chaining) や OpenDaylight についての調査、主に Openstack tacker を使っての環境構築、検証に携わる。

●スタッフ
- 田中 佑佳 (表紙デザイン、紙面レイアウト)
- 鈴木 教之 (Web 連載編集)

本書のご感想をぜひお寄せください
http://book.impress.co.jp/books/1116101064

アンケート回答者の中から、抽選で商品券（1万円分）や図書カード（1,000円分）などを毎月プレゼント。
当選は商品の発送をもって代えさせていただきます。

● 本書の内容に関するご質問は、書名・ISBN・お名前・電話番号と、該当するページや具体的な質問内容、お使いの動作環境などを明記のうえ、インプレスカスタマーセンターまでメールまたは封書にてお問い合わせください。電話やFAX等でのご質問には対応しておりません。なお、本書の範囲を超える質問に関しましてはお答えできませんのでご了承ください。

● 落丁・乱丁本はお手数ですがインプレスカスタマーセンターまでお送りください。送料弊社負担にてお取り替えさせていただきます。但し、古書店で購入されたものについてはお取り替えできません。

■読者の窓口
インプレスカスタマーセンター
〒101-0051 東京都千代田区神田神保町一丁目105番地
TEL 03-6837-5016　／　FAX 03-6837-5023
info@impress.co.jp

■書店／販売店のご注文窓口
株式会社インプレス 受注センター
TEL 048-449-8040
FAX 048-449-8041

OpenDaylight構築実践ガイド（Think IT Books）

2016年10月1日　初版発行

著　者　倉橋 良、鳥居 隆史、安座間 勇二、高橋 信行
発行人　土田 米一
編集人　高橋 隆志
発行所　株式会社インプレス
　　　　〒101-0051　東京都千代田区神田神保町一丁目105番地
　　　　TEL　03-6837-4635（出版営業統括部）
　　　　ホームページ　http://book.impress.co.jp/

本書は著作権法上の保護を受けています。本書の一部あるいは全部について（ソフトウェア及びプログラムを含む）、株式会社インプレスから文書による許諾を得ずに、いかなる方法においても無断で複写、複製することは禁じられています。

Copyright © 2016 Ryo Kurahashi, Takashi Torii, Yuji Azama, Nobuyuki Takahashi. All rights reserved.

印刷所　京葉流通倉庫株式会社
ISBN978-4-8443-8165-5　C3055
Printed in Japan